8017

UNIVERSITY SERIES IN MODERN ENGINEERING

UNIVERSITY SERIES IN MODERN ENGINEERING

SYSTEMS & SIGNALS
N. Levan
1983, x + 173 pp.
ISBN 0-911575-25-1 Optimization Software, Inc.
ISBN 0-387-90900-1 Springer-Verlag New York Berlin Heidelberg Tokyo
ISBN 3-540-90900-1 Springer-Verlag Berlin Heidelberg New York Tokyo

ELEMENTS OF STATE SPACE THEORY OF SYSTEMS
A.V. Balakrishnan
1983, vii + 187 pp.
ISBN 0-911575-27-8 Optimization Software, Inc.
ISBN 0-387-90904-4 Springer-Verlag New York Berlin Heidelberg Tokyo
ISBN 3-540-90904-4 Springer-Verlag Berlin Heidelberg New York Tokyo

KALMAN FILTERING THEORY
A.V. Balakrishnan
1984, xii + 222 pp.
ISBN 0-911575-26-X Optimization Software, Inc.
ISBN 0-387-90903-6 Springer-Verlag New York Berlin Heidelberg Tokyo
ISBN 3-540-90903-6 Springer-Verlag Berlin Heidelberg New York Tokyo

A.V. BALAKRISHNAN

KALMAN FILTERING THEORY

OPTIMIZATION SOFTWARE, INC.
PUBLICATIONS DIVISION, NEW YORK

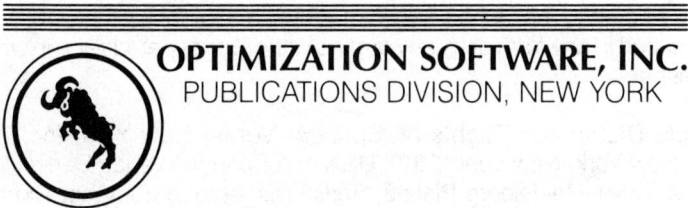

Author
A.V. Balakrishnan
School of Engineering
University of California
Los Angeles, California 90024
USA

Library of Congress Cataloging in Publication Data

Balakrishnan, A.V.
Kalman filtering theory.

(University series in modern engineering)
Bibliography: p.
Includes index.
1. Control theory. 2. Estimation theory. 3. Kalman filtering. I. Title. II. Series.
QA402.3.B287 1984 003 83-25122
ISBN 0-911575-26-X

Worldwide Distribution Rights by Springer-Verlag New York, Inc., 175 Fifth Avenue, New York, New York 10010, USA and Springer-Verlag Berlin Heidelberg New York Tokyo, Heidelberg Platz 3, Berlin-Wilmersdorf 33, West Germany.

ERRATA

page	line	as was	change to
Preface	+5	not not	not
12	-1	converge as $n \to \infty$.	converge as $n \to \infty$, even though bounded.
73	+8	$(P - P_c)$	$(P_c - P)$
	-5	$P_0 \geq \Lambda$;	$P_0 \leq \Lambda$;
	-3	$P_0 \leq \Lambda$.	$P_0 \geq \Lambda$.
79	+9	is white also. It is..	is white also. We assume furthermore that the signal and noise processes are mutually independent; or, equivalently that the white noise processes N_n^S and N_n^O are mutually independent. It is ...
80	+1	We shall ...	-- (delete)
	+2	$F_n G_m^*$ ...	-- (delete)
	+3	implying ...	-- (delete)
	+4	cesses. Another...	Another...
89	-7	$+ K(C_n H_{n-1}$ ...	$+ K_n(C_n H_{n-1}$ ...
90	-1	$(v_n - (A_{n-1}$...	$(v_n - C_n(A_{n-1}$...
92	+9	v_{n-1}	$v_{n-1} + U_{n-1}$
95	-2	$\hat{x}_{n-1} + B_{n-1}U_{n-1}$...	$\hat{x}_{n-1} + U_{n-1}$...
97	-11	$C_n R_n C_n^*(I + C_n R_n$...	$C_n R_n C_n^*(G_n G_n^* + C_n R_n$...
98	+10	$v_n - C_n x_n$	$v_n - m_n$
100	+12	equal to the var-	-- (delete)
	+13	iance of the innovation and is	-- (delete)
101	all		replace with p. 102
102	all		replace with p. 101

Please turn over

page	line	as was	change to
original 102	-9	$E[x_0 x_0^*]$	$E[\tilde{x}_0 \tilde{x}_0^*]$
104	-9	x_n.	$\hat{x}_n$.
	-4	$E[[x_0, x]^2]\ldots$	$\geq E[[x_0, x]^2]\ldots$
105	-4	$E[(x_n - x_n^a)(x_n - x_n^a)^*]$	$E[(x_n - \hat{x}_n^a)(x_n - \hat{x}_n^a)^*]$
108	+2	$\frac{d}{d\lambda}\ \Phi(I + H(\ldots$	$\frac{d}{d\lambda}\ (I + H(\ldots$
109	+8	$A^{*k} C^k v_{m+k-n}$.	$A^{*k} C^* v_{m+k-n}$.
117	-2	$\ldots C^* C)^{-1}$.	$\ldots C^* C)^{-1} P_2$.
132	+8	Show that $(A \sim F)$ is not Controllable if P	-- (delete)
	+9	is singular.	-- (delete)
170	+12	a^N	a^n
	-3	v_n	v_N
216	+7	$(G_k G_k^*)(I-(G_k G_k^*)^{-1} C_k P_k C_k^*)^{-1}$	$(I + C_k H_{k-1} C_k^* (G_k G_k^*)^{-1})$ $\times\ (G_k G_k^* - C_k P_k C_k^*)$ $\times\ (I + (G_k G_k^*)^{-1} C_k H_{k-1} C_k^*)$

ABOUT THE AUTHOR

Professor A. V. Balakrishnan is the founding Chairman of the Department of System Science in the School of Engineering and Applied Science at UCLA. He is a Fellow of the IEEE, and past Chairman of the IEEE Information Theory Group. He has received a Certificate of Recognition from the National Aeronautics and Space Administration for his scientific contributions to the processing of Flight Test data. More recently, he was awarded the Guillemin Prize for his many contributions to Communication and Control Systems Engineering. He is also the recipient of the Silver Core Award from the International Federation of Information Processing. He is the author of several books and monographs in Applied Mathematics and Engineering, and serves on the Editorial Boards of numerous journals and monograph series.

CONTENTS

PREFACE

This is a textbook intended for a one-quarter (or one-semester, depending on the pace) course at the graduate level in Engineering. The prerequisites are Elementary State-Space theory and Elementary (second-order Gaussian) Stochastic Process theory. As a textbook, it does not not purport to be a compendium of all known work on the subject. Neither is it a "trade book." Rather it attempts a logically sequenced set of topics of proven pedagogical value, emphasizing theory while not devoid of practical utility. The organization is based on experience gained over a period of ten years of classroom teaching. It develops those aspects of Kalman Filtering lore which can be given a firm mathematical basis, avoiding the industry syndrome manifest in professional short courses: "Here is the recipe. Use it, it will "work"!"

The first two chapters cover review material on State-Space theory and Signal (Random Process) theory -- necessary but not sufficient for the sequel. The third chapter deals with Statistical Estimation theory, the mathematical framework on which Kalman Filtering rests. The main chapter is the fourth chapter dealing with the subject matter per se. It begins in Section 4.1, with the basic theory and formulas, making a compromise in generality between too many obscuring details and too little practical application. Thus we consider only the case where the observation noise is white and is independent of the signal, although we allow the system to be time-varying. Because of the uncertainty in the initial covariances, in practice no Kalman filter can be optimal except in the steady state -- and this is by far its important use. Hence Section 4.2 specializes to time-invariant systems and considers asymptotic behavior of the filter. Section 4.3 examines the steady-state results from the frequency-domain point of view, relating them to the more classical transfer-function approach. In Section 4.4 we study a canonical application of Kalman filtering: to System Identification. In Section 4.5 we study the "Kalman smoother": the on-line version of two-sided interpolation. In Sections 4.6 and 4.7 we study generalizations of the basic theory of Section 4.1; thus we allow the signal and noise to be correlated in Section 4.6, and allow the observation noise to be non-white in Section 4.7. Section 4.8 features a simple example which illustrates some of the theory and techniques discussed in the chapter.

The book concludes with a chapter on Likelihood Ratios in which the Kalman filter formulation plays an essential role.

We only consider discrete-time models throughout, since all Kalman filter implementation envisaged involves digital computation.

The problems accompanying each chapter serve the traditional role of testing the student's comprehension of the text, with an occasional foray into areas of contiguous interest.

NOTATION

Square Matrices

I ~ Identity Matrix

Tr. = Trace

$|A|$ Determinant of A

Rectangular Matrices

A^* = (Conjugate) Transpose of A

$[A, B]$ = Tr. AB^*

$\|A\| = \sqrt{[A, A]}$

$\{a_{ij}\}$ ~ Matrix with entries a_{ij}

'Column' vector v ~ v^*v is 1×1

'Row' vector c ~ cc^* is 1×1

Self-adjoint Matrices

A Self-adjoint ~ $A = A^*$

A Nonnegative ~ $[Ax, x] \geqslant 0$ for every x

$A \geqslant B \longleftrightarrow (A-B)$ is nonnegative definite

Gradient of a Function

If $g(\theta)$ is a scalar function of θ:

Gradient of $g(\theta) = \nabla_\theta g(\theta)$,

where

$$(\nabla_\theta g(\theta))h = \frac{d}{d\lambda} g(\theta + \lambda h)\Big|_{\lambda = 0}$$

$\nabla_\theta g(\theta)$ is $1 \times m$ if θ is $m \times 1$

Random Variables

$E(\cdot)$ ~ Expected Value

$E(\cdot|\cdot)$ ~ Conditional Expectation

$p(\cdot)$ ~ Probability Density Function

Chapter 1.

REVIEW OF LINEAR SYSTEM THEORY

A Kalman filter is a linear system. This chapter presents a brief review of Linear Systems theory from the "state-space" point of view, since the Kalman filter is best described in that way. For an introductory treatment of State Space theory, the reader is referred to [2]. More advanced treatments may be found in [7], [12], [20], among other texts.

A system is characterized by its "input," its "state" and the "output." These are functions of time. Time may be continuous or discrete -- in the latter case time is indexed by the integers. We shall only be concerned with the discrete case in this book.

Let $\{u_n\}$ denote the input and $\{v_n\}$ the output. A linear system for our purposes is then completely characterized by a "state-input" equation

$$x_{n+1} = A_n x_n + B_n u_n \tag{1.1}$$

and by an "output-state-input" equation

$$v_n = C_n x_n + D_n u_n \quad , \qquad (1.2)$$

where A_n is a square matrix and B_n, C_n, D_n are rectangular matrices. If the state-space dimension is p, then A_n will be $p \times p$. If the input sequence is such that each u_n is $q \times 1$, then B_n will be $p \times q$. If the output sequence is such that each v_n is $m \times 1$, then C_n will be $m \times p$ and D_n will be $m \times q$. We can "solve" (1.1), (1.2) or express the output in terms of the state at some initial time and the input. Thus we have, taking the initial (or starting) time to be k:

$$x_n = \psi_{n,k} x_k + \sum_{i=k}^{n-1} \psi_{n,i+1} B_i u_i \quad , \qquad (1.3)$$

where $\psi_{n,k}$, called the State-Transition Matrix, is defined by

$$\psi_{n,k} = A_{n-1} \cdots A_k \quad , \qquad k \leq n-1 \quad ; \qquad (1.4)$$

$$\psi_{n,n} = I \qquad \text{(Identity Matrix)}^{\dagger} \; .$$

Note that it has the "transition" property:

$$\psi_{n,k} \psi_{k,m} = \psi_{n,m} \; . \qquad (1.5)$$

From (1.3) which specifies the state at any time n, $n \geq k$, the output is readily expressed explicitly in terms of the "initial" state and the current input as:

† Here and throughout, the letter I will always denote the identity matrix regardless of dimension.

$$v_n = C_n \psi_{n,k} x_k + \sum_{i=k}^{n-1} C_n \psi_{n,i+1} B_i u_i + D_n u_n . \qquad (1.6)$$

Here the first term is the "initial state" (or initial condition) response and the second term is the "input response." The function

$$\begin{aligned} W_{n,i} &= C_n \psi_{n,i+1} B_i , & i < n , \qquad (1.7) \\ &= D_n & i = n , \end{aligned}$$

is referred to as the "weighting matrix" or "weighting pattern" of the system.

Time Invariant Systems

We are most concerned with the case where the system is the "time invariant," where the system matrices are all independent of time:

$$A_n = A ,$$

$$B_n = B ,$$

$$C_n = C ,$$

$$D_n = D ,$$

so that (1.1), (1.2) become:

$$\left. \begin{aligned} x_{n+1} &= Ax_n + Bu_n \\ v_n &= Cx_n + Du_n \end{aligned} \right\} . \qquad (1.8)$$

In this case the state-transition matrix

$$\psi_{n,k} = A^{n-k}$$

and hence

$$x_n = A^{n-k} x_k + \sum_{i=k}^{n-1} A^{n-i-1} Bu_i \tag{1.9}$$

and the output

$$v_n = CA^{n-k} x_k + \sum_{i=k}^{n-1} CA^{n-i-1} Bu_i + Du_n . \tag{1.10}$$

Note the "time-invariance" property: the response (state or output) is invariant with respect to any time translation. In particular, it is customary to set the initial time to zero ($k = 0$ in (1.9), (1.10)). The system weighting pattern depends also only on the time difference:

$$\begin{aligned} W_{n,i} &= CA^{n-1-i} B , & n > i , \\ &= D & n = i . \end{aligned}$$

It is more convenient now to write

$$W_k = CA^k B , \qquad k \geq 0 . \tag{1.11}$$

Then (1.10) becomes:

$$v_n = CA^n x_0 + \sum_0^{n-1} W_{n-1-i} u_i + Du_n , \qquad n \geq 0 .$$

We can combine the second and third terms and write

$$v_n = CA^n x_0 + \sum_0^n w_i u_{n-i} \tag{1.10a}$$

by defining

$$w_i = W_{i-1} \quad , \qquad i \geq 1 \quad ,$$
$$\quad = D \quad , \qquad i = 0 \quad .$$

Of the many descriptive properties of time invariant systems defined by (1.8) we are mainly interested in three. These are:

STABILITY

CONTROLLABILITY

OBSERVABILITY .

Let us discuss these in turn.

Stability

We shall say that a state x ($x \in R^p$, the linear space of $p \times 1$ matrices) is stable (or "A-stable") if

$$\lim_n A^n x = 0 \quad . \tag{1.12}$$

In reference to (1.9), this means that asymptotically the initial conditions term therein will vanish, and similarly also in (1.10). The class of all stable states is a linear subspace. We call this the stable subspace. A system is stable if all its states are stable. A necessary and sufficient condition for a system to be stable is that the eigenvalues of A be all (strictly) less than 1 in magnitude. In that case we also say that the the matrix A is stable. We note that A is stable if A^*A is stable, although not conversely.

For a stable system, the weighting pattern of the system can be characterized in terms of its Fourier transform -- called the input-output "transfer function":

$$\psi(\lambda) = \sum_0^\infty w_k \, e^{2\pi i \lambda k} \,, \qquad -\tfrac{1}{2} < \lambda < \tfrac{1}{2} \,, \tag{1.13}$$

$$= D + e^{2\pi i\lambda} \, C \sum_0^\infty A^k \, e^{2\pi i k\lambda} \, B \,.$$

Let z be a complex variable. Then

$$\sum_0^\infty w_k \, z^k$$

converges for $|z| \leq 1$ and is called the z-transform. From (1.11) we have readily that

$$\sum_0^\infty w_k \, z^k = zC(I - zA)^{-1}B + D \,, \qquad |z| < 1 \,. \tag{1.14}$$

All the properties of the z-transform can be inferred from the transfer function:

$$\psi(\lambda) = C(I - e^{2\pi i\lambda}A)^{-1} \, B \, e^{2\pi i\lambda} + D \,, \qquad -\tfrac{1}{2} < \lambda < \tfrac{1}{2} \,. \tag{1.15}$$

(It is of interest to mention in this connection a problem of importance, even if beyond our scope: given a system transfer function $\psi(\lambda)$, when can we express it in the form (1.15), for appropriate A, B, C?) Note that the z-transform (1.14) continues to be defined even if A is not stable, for $|z| < r$ for some r, $0 < r \leq 1$; and hence its importance.

Controllability

We say that a state x is Controllable (reachable is a better word) if it can be reached from the zero state in some finite number of steps by an appropriate input. More precisely, x is "Controllable" if for some n and some $\{u_i\}$,

$$x_0 = 0 ,$$

$$x_{k+1} = Ax_k + Bu_k , \qquad 0 \leq k \leq n-1 ,$$

$$x_n = x ;$$

(or, equivalently, in terms of the explicit expression (1.9) for the state any time,

$$x = \sum_0^{n-1} A^{n-1-k} Bu_k). \qquad (1.16)$$

The controllable states form a linear subspace which we label the "Controllable Subspace." When the latter is the whole state space then we say that the state space is Controllable. We shall then abbreviate this to: $(A \sim B)$ is Controllable.

A necessary and sufficient condition for Controllability is that the (compound) "Controllability Matrix"

$$| \; B \quad AB \cdots A^{p-1}B \; | \qquad (1.17)$$

has full rank. Or, equivalently, the $p \times p$ matrix

$$R_C = \sum_0^{p-1} A^j BB^* A^{*j}$$

is nonsingular. (Recall that A is $p \times p$.) Moreover, the subspace of controllable states is precisely the range of R_C.

If the state space is controllable, we can express the output v_n entirely in terms of the input history. To begin with, we have

$$x_0 = \sum_0^{k-1} A^j B \, u_{k-1-j}$$

for some k, and we can rewrite this reversing time as:

$$x_0 = \sum_{-k}^{-1} A^{-1-j} \, B\tilde{u}_j \quad ,$$

where

$$\tilde{u}_n = u_{k+j} \quad , \qquad -k \le j \le -1 \quad .$$

In other words, we may think of the initial state x_0 being accounted for in terms of an appropriate input history. Moreover, we can then write the current state x_n in the form:

$$\begin{aligned} x_n &= A^n \sum_{-k}^{-1} A^{-1-j} \, Bu_j + \sum_0^{n-1} A^{n-1-i} \, Bu_i \quad , \\ &= \sum_{-k}^{n-1} A^{n-1-i} \, B\bar{u}_i \quad , \end{aligned}$$

where

$$\begin{aligned} \bar{u}_i &= u_i \quad , \qquad 0 < i < n-1 \quad , \\ &= \tilde{u}_i \quad , \qquad -k < i \le -1 \quad . \end{aligned}$$

We can further rewrite this in the "generic" form:

$$x_n = \sum_{-\infty}^{n-1} A^{n-1-i} Bu_i ,$$

where the input sequence u_i is zero for $-\infty < i \leq -N$. In turn, we can express the output as:

$$\begin{aligned} v_n &= \sum_{-\infty}^{n-1} W_{n-1-i} u_i + Du_n , \\ &= \sum_{0}^{\infty} W_j u_{n-1-j} + Du_n , \\ &= \sum_{0}^{\infty} w_j u_{n-j} . \end{aligned} \qquad (1.18)$$

In other words, the output is expressed entirely in terms of the input without introducing state, as a consequence of controllability. Moreover, it is possible (although it is beyond our present scope) to deduce the "state-space" description in which the state space is controllable, starting from (1.18).

Remark. From (1.10a) we note that if the system is stable, the first term in (1.10a) goes to zero for large n, so that the representation (1.18) holds "asymptotically" for stable systems.

Observability

To introduce the notion of Observability, let us begin with a problem -- one which is not without practical importance. Let us assume that the system is known: A, B, C and

D are given. We are also given a sequence of input-output pairs: (u_i, v_i), $i = 1,\ldots,n$, u_i being the input and v_i the output. Can we determine the corresponding states $x_1, \ldots, x_n$ from this data? To answer this question we may proceed as follows. Since

$$v_i = CA^{i-1}x_1 + \sum_{1}^{i-1} CA^{i-1-k} Bu_k , \qquad i = 1,\ldots,n ,$$

let us subtract the response to the known input and define

$$\tilde{v}_i = v_i - \sum_{1}^{i-1} CA^{i-1-k} Bu_k .$$

Then we have

$$\tilde{v}_i = CA^{i-1}x_1 , \qquad i = 1,\ldots,n . \tag{1.19}$$

If we can determine x_1 from this, then of course

$$x_i = A^{i-1}x_1 + \sum_{1}^{i-1} A^{i-1-k} Bu_k$$

will determine the succeeding states for us. Now we may regard (1.19) as a set of n "equations" to solve for x_1. Moreover, the equations being linear, we see that (1.19) has a unique solution only if the homogeneous equation

$$0 = CA^{i-1}z , \qquad i = 1,\ldots,n , \tag{1.20}$$

has no "nonzero" solution z. Recalling that the state space is of dimension p and that A is $p \times p$, we note that

$$0 = CA^{i-1}z , \qquad i = 1,\ldots,p , \tag{1.21}$$

implies (1.20) for $n \geq p$.

We now define: a state x is <u>Unobservable</u> if

$$CA^k x = 0 \quad , \qquad k \geq 0 \quad . \tag{1.22}$$

The class of unobservable states is clearly a linear subspace. Its orthogonal complement is called the Observable Subspace. We say that the state space is Observable if the subspace of unobservable states contains only the zero state.

Let p be the dimension of the state space. Then the state space is observable if and only if

$$R_0 = \sum_0^{p-1} A^{*j} C^* C A^j$$

is nonsingular. Or, equivalently, the (compound) matrix

$$\begin{vmatrix} C \\ CA \\ \vdots \\ CA^{p-1} \end{vmatrix}$$

is nonsingular. Moreover, if the state space is observable, we can, going back to (1.19), determine x_1 as:

$$x_1 = R_0^{-1} \sum_1^p A^{*i-1} C^* \tilde{v}_i \quad . \tag{1.23}$$

We also use the notation "$(C \sim A)$ Observable" to denote that the state space is observable. Note, in particular, that "$(C \sim A)$ Observable" is equivalent to "$(A^* \sim C^*)$ Controllable."

★ PROBLEMS ★

<u>Problem 1.1</u>

Suppose $(C \sim A)$ is observable. If $(I - CK)$ is nonsingular, then $(C \sim (I - KC)A)$ is also observable.

<u>Hint</u>:

$$C(I - KC)Ax = 0 \quad => \quad (I - CK)CAx = 0$$

$$=> \quad CAx = 0 \quad ,$$

hence

$$C(I - KC)A(I - KC)Ax \quad => 0 \quad ,$$

$$=> \quad C(I - KC(A^2x = 0 \quad ,$$

$$=> \quad (I - CK)CA^2x = 0 \quad ,$$

$$=> \quad CA^2x = 0 \quad ,$$

etc.

<u>Problem 1.2</u>

Suppose A is nonsingular. Show that $(C \sim A^{-1})$ is observable if $(C \sim A)$ is.

<u>Problem 1.3</u>

Show that $(C \sim A)$ Observability is equivalent to:

$(C^*C \sim A)$ Observability

$(\sqrt{C^*C} \sim A)$ Observability .

<u>Problem 1.4</u>

Construct a (square) matrix A such that for some x

$$\|A^n x\|$$

does <u>not</u> converge as $n \to \infty$.

Chapter 2.

REVIEW OF SIGNAL THEORY

In this chapter we present a brief review of the salient facts about Signals essential in the sequel. For more details, including details of proofs, the references [13, 14, 17, 18, 19] may be consulted.

As in the case of Systems in Chapter 1, we shall consider only the "discrete time" case where the independent variable may then be replaced by integers. Thus we shall use the notation: s_n for the nth sample, counting from some (arbitrary) initial sample (time). (The samples need not necessarily be taken at some fixed rate, although that would be the most common situation.) We take each s_n to be an $m \times 1$ column vector. Since the initial starting time is arbitrary, we may allow n to be positive as well as negative as need arises. Although it would be impossible in any physical device to process a non-finite number of samples, it would be equally unrealistic to limit the number of samples to be a fixed finite number, fixed once and for all. Hence we idealize our signals as nonterminating sequences: $\{s_n\}$, $n = 1,2,\ldots,$ running through all the positive integers; or, as necessary, the negative integers as well.

Spectral Theory of Signals with Finite Power

We say a signal sequence† $\{s_n\}$ has finite energy if

$$\lim_{N\to\infty} \sum_{-N}^{N} \|s_n\|^2 < \infty .$$

We shall have little to do with the associated theory.

We say that a signal $\{s_n\}$ has "finite power"

$$\lim_{N\to\infty} \frac{1}{2N} \sum_{-N}^{N} s_n s_n^* = P_s < \infty . \qquad (2.1)$$

Actually we shall demand a little bit more than (2.1). (This is usually assumed implicitly in the engineering literature.) Thus, following Wiener's definitive work [18], we shall assume that

$$\lim_{N\to\infty} \frac{1}{2N} \sum_{-N}^{N} s_n s_{n+m}^* = R_m$$

exists and is finite for each $m \geq 0$. This does NOT necessarily follow from (2.1)! Note that

† In what follows we omit the qualification "sequence."

$$R_m^* = \lim_N \frac{1}{2N} \sum_{-N}^{N} s_{n+m} s_n^*$$

$$= \lim_N \frac{1}{2N} \sum_{-N}^{N} s_n s_{n-m}^*$$

$$= R_{-m} \quad . \qquad (2.2)$$

Of course,

$$R_0 = P_s \quad \text{(signal covariance)} \quad .$$

A typical example of a signal with finite power is:

$$s_n = a \cos(2\pi n \lambda_o + \theta) \quad , \qquad -\tfrac{1}{2} < \lambda_o < \tfrac{1}{2} \quad ,$$

which is periodic if λ_o is rational. Note in this case:

$$R_m = aa^* \lim \frac{1}{2N} \sum_{-N}^{N} \cos(2\pi n \lambda_o + \theta) \cdot \cos(2\pi(n+m)\lambda_o + \theta)$$

$$= (aa^*)\tfrac{1}{2} \cos 2\pi m \lambda_o \quad , \qquad \lambda_o \neq 0 \quad .$$

Note that

$$R_m = \int_{-\frac{1}{2}}^{\frac{1}{2}} e^{2\pi i \lambda m} \left(\frac{aa^*}{4}\right)(\delta(\lambda-\lambda_o) + \delta(\lambda+\lambda_o))\, d\lambda$$

and, for $\lambda_o \neq 0$, is the same whatever the "phase angle" θ. This result can be generalized. We have in fact the "spectral representation" theorem due to Bochner-Khinchin-Wiener:

$$R_m = \int_{-\frac{1}{2}}^{\frac{1}{2}} e^{2\pi i \lambda m}\, p_s(\lambda)\, d\lambda \quad , \qquad (2.3)$$

where $p_s(\lambda)$ is self-adjoint, nonnegative definite (and of course may contain delta-functions) and is known as the spectral density of the signal. In fact, it may be obtained by the Fourier series expansion:

$$p_s(\lambda) = \sum_{-\infty}^{\infty} e^{-2\pi i\lambda m} R_m \quad , \qquad (2.4)$$

or $\{R_m\}$ are the Fourier coefficients of $p_s(\lambda)$. From (2.4) we have that

$$p_s(\lambda)^* = p_s(+\lambda) \quad .$$

Let such a signal $\{s_n\}$ be the input to a linear system with transfer function $\psi(\lambda)$. Let $\{v_n\}$ denote the corresponding output, so that

$$v_n = \sum_0^{\infty} w_k s_{n-k} = \sum_{-\infty}^{n} w_{n-k} s_k \quad ,$$

$$\psi(\lambda) = \sum_0^{\infty} w_k \, e^{2\pi ik\lambda} \quad , \qquad \sum_0^{\infty} ||w_k||^2 < \infty \; .$$

Then

$$\frac{1}{2N} \sum_{-N}^{N} v_n v_{n+m}^* = \sum_0^{\infty} w_k \left(\frac{1}{2N} \sum_{-N}^{N} s_{n-k} s_{n+m-j}^* \right) w_j^* \quad ;$$

and upon taking limits as $N \to \infty$, we obtain

$$R_m^v = \lim_{N\to\infty} \frac{1}{2N} \sum_{-N}^{N} v_n v_{n+m}^*$$

$$= \sum_0^\infty \sum_0^\infty w_k \; R_{m+k-j} \; w_j^*$$

$$= \int_{-\frac{1}{2}}^{\frac{1}{2}} e^{2\pi i\lambda m} \left(\sum_0^\infty \sum_0^\infty w_k \; e^{2\pi i\lambda k} \; p_s(\lambda) \; e^{-2\pi i\lambda j} \; w_j^* \right)$$

$$= \int_{-\frac{1}{2}}^{\frac{1}{2}} e^{2\pi i\lambda m} \; \psi(\lambda) \; p_s(\lambda) \; \psi(\lambda)^* \; d\lambda$$

(where * indicates "conjugate-transpose"!). In other words, the output $\{v_n\}$ has "time average" properties similar to that of the input $\{s_n\}$, and the spectral density of the output is

$$p_v(\lambda) = \psi(\lambda) \; p_s(\lambda) \; \psi(\lambda)^* \quad . \tag{2.5}$$

In the special case where the system has the structure:

$$v_n = Cx_n \tag{2.6}$$

$$x_{n+1} = Ax_n + Fs_n \quad ,$$

we see (assuming A is stable) that

$$v_n = \sum_0^\infty CA^k F \; s_{n-1-k} \quad ,$$

where now

$$\psi(\lambda) = \sum_0^\infty CA^k \; e^{2\pi i\lambda k} \; F$$

$$= C(I - Ae^{2\pi i\lambda})^{-1} \; F \quad ;$$

so that we have for the spectral density:

$$p_v(\lambda) = C(I - Ae^{2\pi i\lambda})^{-1} F p_s(\lambda) F^* (I - A^* e^{-2\pi i\lambda})^{-1} C^* . \tag{2.7}$$

Stochastic Signals: Second Order Theory

In his pioneering work [18], Wiener used the theory sketched above for the description of the signals. This theory is a "steady state" theory. It is possible to obtain a more general theory which enables us to include the "transient" or "nonsteady state" analysis and at the same time make the steady state analysis easier. This is accomplished by the introduction of stochastic signals, including a "signal generation" theory, which enables us to construct a signal with given spectral density.

A stochastic signal for us is a random process (sequence) (or time series) -- a sequence of random variables whose joint distributions of any order are given (or calculable, in principle). A Gaussian stochastic signal is one whose joint distributions are all Gaussian. We shall only need to consider Gaussian signals in Kalman filtering theory. More generally, in linear filtering theory we shall be concerned only with moments up to the second order (means and covariences); and hence we shall need only "second order" theory. On the other hand, we may replace the given process by a Gaussian with the same means and covariances, and hence we may as well consider only Gaussian processes.

We say that a stochastic signal $\{s_n\}$ is (second order) stationary if $E[s_n]$ is independent of n (where here and below $E[\cdot]$ denotes expectation or "phase average") and the covariance

$$E[(s_n - E[s_n])(s_{n+m}^* - E[s_{n+m}^*])] = R_m .$$

In other words, a "time translation" does not make any difference. In particular, if the process is Gaussian, then the density functions are also invariant with respect to a "time shift":

$$p(s_n, s_{n+1}, \ldots, s_{n+p}) = p(s_{n+m}, s_{n+m+1}, \ldots, s_{n+p+m})$$

for all n, m and p. For a second order stationary process we have again the Wiener-Khinchin-Bochner theorem (cf. [14]):

$$R_m = \int_{-\frac{1}{2}}^{\frac{1}{2}} e^{2\pi i m\lambda} \, p(\lambda) \, d\lambda ,$$

where $p(\lambda)$ is the "spectral density" of the process (and may contain δ-functions, or more strictly speaking:

$$R_m = \int_{-\frac{1}{2}}^{\frac{1}{2}} e^{2\pi i m\lambda} \, dP(\lambda) ,$$

where $P(\lambda)$ is called the spectral distribution). In the case

$$dP(\lambda) = p(\lambda) \, d\lambda$$

(i.e., the process has a spectral density) it may be determined by the Fourier series:

$$p(\lambda) = \sum_{-\infty}^{\infty} e^{-2\pi i m\lambda} R_m \quad .$$

Note that $p(\lambda)$ is self-adjoint and nonnegative definite. Also

$$p(-\lambda) = \sum_{-\infty}^{\infty} e^{-2\pi i m\lambda} R_m^* \quad ,$$

so that

$$p_{ii}(\lambda) = p_{ii}(-\lambda) \geq 0 \quad ,$$

where $p_{ii}(\lambda)$ denotes the diagonal terms.

The problem of estimating $p(\lambda)$ from one "long" realization of the process is referred to as "spectral estimation" and is an important one, although beyond the scope of the present work.

Let us observe now that the spectral theory of signals with finite power that we started with, shows strong resemblance to the spectral theory of stochastic signals. We can make this connection more precise by invoking the central result of ergodic theory: For a stationary Gaussian process which has a continuous spectral distribution function, the "phase average" may be replaced by the time average. Thus, if there are no "jumps" in the spectral distributions (or no "delta functions" in the spectral density, in engineering language), then

$$E[f(s_n)] = \lim_{N\to\infty} \frac{1}{2N} \sum_{-N}^{N} f(s_k)$$

so that in particular

$$E[s_n] = \lim_{N\to\infty} \frac{1}{2N} \sum_{-N}^{N} s_k \ .$$

Also:

$$E[s_n s_{n+m}^*] = \lim_{N\to\infty} \frac{1}{2N} \sum_{-N}^{N} s_k s_{k+m}^* \ .$$

In particular, therefore

$$R_m = E[s_n s_{n+m}^*] = \lim_{N\to\infty} \frac{1}{2N} \sum_{-N}^{N} s_k s_{k+m}^* \ .$$

Thus we may consider signals as ergodic Gaussian processes, so that we retain the "time average" notion as well as the phase average or statistical average notion. In this sense, modeling signals as stationary Gaussian signals is more general, and is the generally accepted modern view.

Let a stationary (zero mean) Gaussian signal $\{s_n\}$ with spectral density $p_s(\lambda)$ be the input to a time invariant linear system and let the output be $\{v_n\}$, so that

$$v_n = \sum_{0}^{\infty} w_k s_{n-k} \ .$$

Then the output $\{v_n\}$ is also Gaussian and stationary with spectral density $p_v(\lambda)$:

$$p_v(\lambda) = \psi(\lambda)\, p_s(\lambda)\, \psi(\lambda)^* \ ,$$

where

$$\psi(\lambda) = \sum_{0}^{\infty} w_k \, e^{2\pi i k\lambda} \qquad \left(\sum_{0}^{\infty} \|w_k\|^2 < \infty \quad \text{is assumed} \right) \ .$$

This can be proved essentially as in the deterministic case, but now using "phase" (statistical) averages:

$$\begin{aligned}
E[v_n v_{n+m}^*] &= \sum_0^\infty \sum_0^\infty w_k \, E[s_{n-k} s_{n+m-j}^*] \, w_j^* \\
&= \sum_0^\infty \sum_0^\infty w_k \int_{-\frac{1}{2}}^{\frac{1}{2}} e^{2\pi i\lambda(m-j+k)} \, p_s(\lambda) \, d\lambda \quad w_j^* \\
&= \int_{-\frac{1}{2}}^{\frac{1}{2}} \psi(\lambda) \, p_s(\lambda) \, \psi(\lambda)^* \, d\lambda \quad .
\end{aligned}$$

In particular, (2.7) holds when the system is specified by (2.6).

<u>Signal Generation Models</u>

A Gaussian signal $\{N_n\}$ such that

$$\begin{aligned}
E[N_n] &= 0 \quad , \\
E[N_n N_m^*] &= 0 \quad , \qquad n \neq m \quad , \\
&= I \quad , \qquad n = m \quad ,
\end{aligned}$$

is called "white noise" with unit covariance. Let

$$\begin{aligned}
s_n &= C_n x_n \quad , \\
x_{n+1} &= A_n x_n + F_n N_n \quad , \quad n \geqslant 0,
\end{aligned} \qquad (2.8)$$

which is then a "model" for generating the signal $\{s_n\}$, referred to as "white noise through a linear dynamic system." It turns out that for all practical purposes this is the only class of signals of interest in filtering theory. Let us now study some properties of the signal process $\{s_n\}$ so generated. We assume that x_0 is Gaussian and is independent of the noise $\{N_n\}$, for $n \geq 0$. Then the process $\{x_n\}$ -- the "state" process -- is Gaussian and Markovian. Thus we can readily verify that

$$E[x_n \mid x_p, x_{p-1}, \ldots, x_1, x_0] = E[x_n \mid x_p] \quad .$$

Indeed, we have only to note that solving (2.8):

$$\begin{aligned} x_n = {} & A_{n-1} \cdots A_p x_p + F_{n-1} N_{n-1} + A_{n-1} F_{n-2} N_{n-2} \\ & + A_{n-1} A_{n-2} F_{n-3} N_{n-3} \\ & + \cdots + A_{n-1} A_{n-2} \cdots A_p F_p N_p \quad , \end{aligned}$$

the noise terms on the right being independent of x_p. Once again, this implies that all the memory resides in the state one step behind.

Let us now calculate the means and covariance of $\{s_n\}$. We have

$$E[s_n] = C_n \, E[x_n]$$

and

$$\begin{aligned} E[x_n] &= A_{n-1} \, E[x_{n-1}] \\ &= A_{n-1} \cdots A_0 E[x_0] \quad . \end{aligned}$$

In particular, s_n has zero mean if the initial state x_0 has zero mean.

Let us next examine the covariance structure. For simplicity, we shall assume that

$$E[x_0] = 0 ,$$

so that

$$E[x_n] = 0 , \qquad n \geq 0 .$$

(Otherwise we may work with "centered" process

$$\tilde{x}_n = x_n - E[x_n]$$

which will satisfy the same dynamics and will have zero mean.)

We have, letting

$$R(m,n) = E[x_m x_n^*] ,$$

that

$$R(n+p,n) = A_{n+p-1} \cdots A_n R(n,n)$$

and $R(n,n)$ satisfies:

$$R(n+1,n+1) = A_n R(n,n) R_n^* + F_n F_n^* ; \qquad R(0,0) = E[x_0 x_0^*] . \tag{2.9}$$

And for the signal covariance we thus have:

$$E[s_m s_n^*] = C_m R(n,m) C_n^* .$$

Of special interest to us is the time invariant case where

$$A_n = A ,$$

$$F_n = F ,$$

$$C_n = C .$$

Then we have:

$$R(m,n) = R(m-n) = A^{m-n} R(m,n) , \qquad m \geq n ,$$

while

$$R(n,n) = A R(n-1,n-1) A^* + FF^* . \qquad (2.10)$$

We can, of course, "solve" this difference equation to obtain

$$R(n,n) = A^n R(0,0) A^{*n} + \sum_0^{n-1} A^k FF^* A^{*k} .$$

Of special interest to us is the case where A is stable. In that case,

$$A^n R(0,0) A^{*n} \to 0$$

and the series

$$\sum_0^{\infty} A^k FF^* A^{k*}$$

converges. Hence

$$\lim_{n\to\infty} R(n,n) = \sum_0^{\infty} A^k FF^* A^{*k} .$$

Let us denote this by R_∞. Then

$$\lim_{n\to\infty} E[x_n x_{n+p}^*] = R_\infty A^{*p} , \qquad p \geq 0 . \qquad (2.11)$$

In other words, the Gaussian process $\{x_n\}$ and hence the signal process $\{s_n = Cx_n\}$ is <u>asymptotically</u> stationary. Note that R_∞ satisfies

$$R_\infty = A R_\infty A^* + FF^* . \qquad (2.12)$$

The spectral density corresponding to the asymptotically

stationary process is

$$p_s(\lambda) = C\, p_x(\lambda)\, C^* \quad , \tag{2.13}$$

where

$$p_x(-\lambda) = \sum_0^\infty e^{-2\pi i\lambda n} A^n R_\infty + R_\infty \sum_0^\infty e^{2\pi i\lambda n} A^{*n} - R_\infty$$

$$= (I - Ae^{-2\pi i\lambda})^{-1} R_\infty + R_\infty (I - A^* e^{2\pi i\lambda})^{-1} - R_\infty$$

$$= (I - Ae^{-2\pi i\lambda})^{-1}$$

$$\cdot\ [R_\infty(I - A^* e^{2\pi i\lambda}) + (I - Ae^{-2\pi i\lambda})R_\infty - (I - Ae^{-2\pi i\lambda})R_\infty(I - A^* e^{2\pi i\lambda})]$$

$$\cdot\ (I - A^* e^{2\pi i\lambda})^{-1} \quad ,$$

where the quantity in square brackets can be evaluated using (2.12), yielding

$$p_x(\lambda) = (I - Ae^{+2\pi i\lambda})^{-1} FF^* (I - A^* e^{-2\pi i\lambda})^{-1}$$

$$= \psi(\lambda)\, \psi(\lambda)^* \quad ,$$

where $\psi(\lambda)$ is the system transfer function

$$\psi(\lambda) = \left(\sum_0^\infty A^n e^{+2\pi i n\lambda} \right) F \quad .$$

In particular, we have the "factorization" for the signal spectral density:

$$p_s(\lambda) = (C\psi(\lambda))(C\psi(\lambda))^* \quad ,$$

where $C\psi(\lambda)$ is the transfer function of a "physically realizable weighting pattern." Finally, let us note that for the stable system (A-stable) we can show that the time average is equal to the (steady state) phase average:

$$\lim_{N\to\infty} \frac{1}{N} \sum_{n=1}^{N} s_{n+p} s_n^* = A^p R_\infty \quad .$$

<u>Example</u>

Let us illustrate our ideas with a simple example. Suppose our stochastic signal arises from sampling periodically the noise response of a linear oscillator. Thus

$$s_n = s(n\Delta) \quad , \tag{2.14}$$

$$\frac{d^2 s}{dt^2} + 2b\frac{ds}{dt} + \omega_0^2 s(t) = N(t) \quad , \tag{2.15}$$

where $N(t)$ is the noise input to the oscillator system and Δ is the sampling interval. We assume that $\omega_0 \Delta \leq \pi$. Our first step is to derive the signal-generation model for (2.14), using (2.15). For this purpose we first rewrite (2.15) in "state-space" form:

$$\left. \begin{aligned} s(t) &= Cx(t) \\ \dot{x}(t) &= Hx(t) + GN(t) \end{aligned} \right\} \quad , \tag{2.16}$$

where

$$H = \begin{bmatrix} 0 & 1 \\ -\omega_0^2 & -2b \end{bmatrix} \quad ,$$

$$C = [\ 1,\ 0\ ] \quad ,$$

$$G = \begin{vmatrix} 0 \\ 1 \end{vmatrix} .$$

Let

$$x_n = x(n\Delta) .$$

Then from (2.16) we can write

$$x_n = e^{H\Delta} x_{n-1} + \zeta_{n-1} , \tag{2.17}$$

where

$$\zeta_{n-1} = \int_0^\Delta e^{H(\Delta-\sigma)} GN(\overline{n-1}\Delta+\sigma) \, d\sigma . \tag{2.18}$$

Let us assume that $N(t)$ is a Gaussian process with mean zero. Then so is ζ_n and further

$$E[\zeta_n \zeta_m^*] = \int_0^\Delta \int_0^\Delta e^{H(\Delta-\sigma)} GR(\overline{m-n}\Delta+\sigma-s)G^* e^{H^*(\Delta-s)} \, d\sigma \, ds ,$$

where

$$R(t_2 - t_1) = E[N(t_1) N(t_2)^*] .$$

We assume that (corresponding to "large bandwidth" noise):

$$R(t) = 0 \quad \text{for} \quad |t| > \Delta ,$$

and further take Δ small enough so that the double integral can be approximated well by

$$= \Delta R(0) FF^* \qquad \text{for} \quad m = n ,$$

$$= 0 \qquad \text{for} \quad m \neq n .$$

Hence we obtain

$$\zeta_n = \sqrt{\Delta R(0)} \, GN_n ,$$

where N_n is white Gaussian with unit variance. We thus have the representation in the canonical form:

$$\left.\begin{aligned} s_n &= Cx_n \\ x_{n+1} &= Ax_n + FN_n \end{aligned}\right\} , \qquad (2.19)$$

where

$$A = e^{H\Delta} ,$$

$$F = \sqrt{\Delta R(0)}\, G .$$

Note that (2.19) is <u>not</u> obtained by directly "discretizing" the differential equation (2.15).

Let us pursue this example further. We assume that

$$\frac{b}{\omega_0} < 1 ,$$

so that we do have an oscillatory system. Then letting

$$\frac{b}{\omega_0} = \xi \qquad \text{(damping ratio)} ,$$

$$\omega_s = \omega_0 \sqrt{1-\xi^2} ,$$

we have

$$A = e^{H\Delta}$$

$$= e^{-b\Delta} \begin{bmatrix} \cos \omega_s\Delta + \dfrac{\xi}{\sqrt{1-\xi^2}} \sin \omega_s\Delta & \dfrac{1}{\omega_s} \sin \omega_s\Delta \\ \dfrac{-\omega_0}{\sqrt{1-\xi^2}} \sin \omega_s\Delta & \cos \omega_s\Delta - \dfrac{\xi}{\sqrt{1-\xi^2}} \sin \omega_s\Delta \end{bmatrix}$$

Note that $(C \sim A)$ is observable. Hence we can obtain a

"difference equation" for s_n. Exploiting the fact that (Cayley-Hamilton Theorem (see [1])):

$$A^2 + (2e^{-b\Delta} \cos \omega_s \Delta)A + e^{-b\Delta} I = 0 ,$$

we obtain

$$s_{n+2} + (2e^{-b\Delta} \cos \omega_s \Delta) s_{n+1} + e^{-b\Delta} s_n$$

$$= \left(e^{-b\Delta} \frac{\sin \omega_s \Delta}{\omega_s} \right) \sqrt{\Delta R(0)}\, N_n .$$

Moreover, A being stable, the signal $\{s_n\}$ is asymptotically stationary and ergodic. Its spectral density is

$$p(\lambda) = C(I - e^{2\pi i\lambda} A)^{-1} FF^* (I - e^{-2\pi i\lambda} A^*)^{-1} C^*$$

$$= \frac{\Delta R(0)\, e^{-2b\Delta}\, \omega_0^2 \sin^2 \omega_s \Delta}{(1-\xi^2)} \cdot \left| e^{-b\Delta} - e^{-2\pi i(\lambda + f_s \Delta)} \right|^{-2}$$

$$\cdot \left| e^{-b\Delta} - e^{-2\pi i(\lambda - f_s \Delta)} \right|^{-2} ,$$

where

$$2\pi f_0 = \omega_0 ; \qquad 2\pi f_s = \omega_s .$$

Note that $p(\lambda)$ has a maximum at

$$\lambda = \pm f_s \Delta \approx \pm f_0 \Delta$$

for small ξ. Also

$$e^{-b\Delta} = e^{-(2\pi f_0 \Delta)\xi}$$

$$\approx 1 ,$$

under our assumptions, for small ξ. Finally,

$$\log p(\lambda) = \text{constant} - \log\left(1 + e^{-2b\Delta} - 2e^{-b\Delta}\cos(\lambda - f_s\Delta)\right) - \log\left(1 + e^{-2b\Delta} - 2e^{-b\Delta}\cos(\lambda + f_s\Delta)\right) ,$$

where the second term is symmetric about $(f_s\Delta)$ with maximum at $(f_s\Delta)$, while the third is symmetric about $(-f_s\Delta)$ with maximum at $(-f_s\Delta)$. The smaller b, the sharper the peaks.

★ PROBLEMS ★

Problem 2.1

Let $\{s_n\}$ be a stationary Gaussian signal with spectral density $p(\lambda)$. Define (for fixed B) the continuous time stochastic process by

$$s(t) = \sum_{-\infty}^{\infty} s_n \frac{\sin \pi(2Bt-n)}{\pi(2Bt-n)} , \qquad -\infty < t < \infty ,$$

where the convergence of the infinite series is taken in the mean square sense. Find the covariance of the process

$$s(k\Delta) , \qquad -\infty < t < \infty ,$$

for fixed Δ, $2B\Delta \leq 1$. Specialize to the case where the spectral density of $\{s_n\}$ is such that

$$p(\lambda) = 1 , \qquad -\tfrac{1}{2} \leq \lambda \leq \tfrac{1}{2} .$$

Hint:

$$E[s(t)\, s(t+k\Delta)^*] = \sum_p R_p \sum_n a_n(t)\, a_{n+p}(t+k\Delta) ,$$

where

$$R_p = E[s_n s_{n+p}^*] \quad ,$$

$$a_n(t) = \frac{\sin \pi(2Bt-n)}{\pi(2Bt-n)} \quad .$$

Hence

$$E[s(t)\, s(t+k\Delta)^*] = \sum_p R_p \frac{\sin \pi(2Bk\Delta-p)}{\pi(2Bk\Delta-p)}$$

$$= \int_{-\frac{1}{2}}^{\frac{1}{2}} \sum_p e^{2\pi i\lambda p} \frac{\sin \pi(2Bk\Delta-p)}{\pi(2Bk\Delta-p)}\, p(\lambda)\, d\lambda$$

$$= \int_{-\frac{1}{2}}^{\frac{1}{2}} e^{2\pi ik\lambda(2B\Delta)}\, p(\lambda)\, d\lambda \quad .$$

Hence spectral density

$$= \frac{1}{2B\Delta}\, p\left(\frac{\lambda}{2B\Delta}\right) \qquad \text{for} \quad -B\Delta < \lambda < B\Delta \quad ,$$

$$= 0 \qquad \text{otherwise} \quad .$$

<u>Problem 2.2</u>

For the signal model:

$$s_n = Cx_n$$

$$x_{n+1} = Ax_n + FN_n \,, \qquad n \geq 0$$

where A is stable and x_0 is independent of the unit-variance white-noise sequence $\{N_n\}$, calculate the mean and variance of the "sample" covariance

$$\left(\frac{1}{N} \sum_{n=1}^{N} s_n s_{n+p}^*\right)$$

for each N. Show that the variance goes to zero as $N \to \infty$.

Problem 2.3

Prove the Schwarz inequality for random vectors x, y:

$$|E[xy^*]|^2 \leq E[\|x\|^2[\ E[\|y\|^2] \quad .$$

Prove the Holder inequality:

$$\sqrt{E[\|x+y\|^2]} \leq \sqrt{E[\|x\|^2]} + \sqrt{E[\|y\|^2]} \quad .$$

Chapter 3.

STATISTICAL ESTIMATION THEORY

The function of a "filter," as the name suggests, is to "filter out" the noise; or equivalently, to extract the desired signal from the corrupting noise. A filter is, in other words, an "estimator" of the signal. The paradigms for this are well-established in statistics -- or, to be more specific, in the statistical theory of estimation. This is the subject matter of the present chapter. It is important not only because it is a logical starting point in developing the Kalman filter theory, but also because it is of independent interest in its own right.

3.1 PARAMETER ESTIMATION: THE CRAMER-RAO BOUND; THE PRINCIPLE OF MAXIMUM LIKELIHOOD

We "observe" an $n \times 1$ vector v. For example, v could be n samples (arranged as a column vector) of a continuous time waveform that is sampled at some discrete intervals of time. Given v we need to estimate an $m \times 1$ "parameter" Θ (see below for examples). We consider the situation where nothing is known about Θ: it is just an "unknown" parameter, and v is modeled as a random variable whose distribution is known for each Θ. We have then what is called a "parameter estimation" problem, in the classical statistical terminology associated with the names of Cramer and Fisher. (See [3]).

Of course, as we shall see quickly, other points of view concerning Θ are possible, although we cannot discuss them all in this book. Which view we adopt is determined by our "track record" of success -- how well we do. We retain the model if our experience is positive and discard it, otherwise.

The unknown parameter Θ is any point in m-dimensional Euclidean space. For our purpose, in this book we need only to consider the case where the distribution of v can be replaced by its density; or, in "pure mathematics" terms, the distribution is absolutely continuous with respect to Lebesgue measure. Thus we are given a family (indexed by Θ) of probability densities:

$$p(v|\theta) \geq 0 \quad ,$$

$$\int_{R^n} p(v|\Theta)\, d|v| = 1 \quad .$$

The notation $p(v|\Theta)$ is intended to suggest that we may think of $p(v|\Theta)$ as the "conditional" density of v given Θ, even though Θ is not a random variable.

Any "estimate" will be (in fact, will have to be) a function of v. Thus, the "hat" denoting an estimate, we can write:

$$\hat{\Theta} = f(v) \quad ,$$

where the function $f(\cdot)$ completely specifies the estimate. How shall we measure the "goodness" of our estimate? Note that $f(v)$ is a random variable. Hence any such measure must involve "statistical averages." We begin with the first moment:

$$E_\Theta(\hat{\Theta}) = \int f(v)\, p(v|\Theta)\, d|v| \quad ,$$

where the subscript Θ indicates that Θ is fixed. The quantity

$$E_\Theta(\hat{\Theta}) - \Theta$$

is called the estimate "bias." The "bias" is clearly a function of Θ:

$$b(\Theta) = E_\Theta(\hat{\Theta}) - \Theta \quad .$$

An estimate is said to be "unbiased" if the bias is zero.

Let Θ_0 denote the "true" value of Θ, which is, of course, unknown and, in fact, unknowable. Nevertheless we can talk about the "error":

$$f(v) - \Theta_0 \quad .$$

Since Θ_0 is unknown, we consider for arbitrary Θ the "mean square error"

$$E_\Theta(\|f(v) - \Theta\|)^2 = \int_{R_n} \|f(v) - \Theta\|^2 \, p(v|\Theta) \, d|v| \quad ,$$

or, more generally, the second moment matrix of the error:

$$\begin{aligned} R(\Theta) &= E_\Theta((f(v) - \Theta)(f(v) - \Theta)^*) \\ &= \int_{R^n} (f(v) - \Theta)(f(v) - \Theta)^* \, p(v|\Theta) \, d|v| \end{aligned} \qquad (3.1.1)$$

as a function of Θ.

The remarkable discovery associated with the names of Cramer and Rao [3, 16] is that one can calculate the "minimal" second moment matrix $R(\Theta)$ -- the minimum of (3.1.1) over the class of all possible estimators $f(\cdot)$ -- without actually needing to know the corresponding estimate (except for its bias). Or, more precisely, we can obtain a lower bound to $R(\Theta)$ in the form

$$R(\Theta) \geq (I + \nabla_\Theta b(\Theta)) \, \Lambda(\Theta)^{-1} \, (I + \nabla_\Theta(b(\Theta))^* \quad , \qquad (3.1.2)$$

where

$$\begin{aligned} \Lambda(\Theta) &= E_\Theta[(\nabla_\Theta \log p(v|\Theta))^*(\nabla_\Theta \log p(v|\Theta))] \\ &= \int_{R^n} (\nabla_\Theta \log p(v|\Theta))^*(\nabla_\Theta \log p(v|\Theta)) \, p(v|\Theta) \, d|v| \quad , \end{aligned}$$

where ∇_Θ denotes the gradient ($1 \times m$ matrix) with respect to Θ. To use (3.1.2) we do need to know the bias in the estimate which can be difficult to evaluate in general. The inequality is most useful, in fact, when the bias is zero. Thus we have that for unbiased estimators the second moment matrix $R(\Theta)$ (which is then also the variance matrix of the "error" $(\Theta - \hat{\Theta})$) has the lower bound ("C-R bound"):

$$R(\Theta) \geq \{E_\Theta[(\nabla_\Theta \log p(v|\Theta))^*(\nabla_\Theta \log p(v|\Theta))]\}^{-1} . \qquad (3.1.3)$$

Again the inequality is most useful when the bound is independent of Θ; its use is circumspect when this is not the case.

Proof of the C-R Bound Formula

Let us see how to prove (3.1.2). First we shall need to assume that

$$\nabla_\Theta \int_{R^n} p(v|\Theta)\, d|v| = \int_{R^n} \nabla_\Theta p(v|\Theta)\, d|v| .$$

Or, in other words, "differentiation with respect to the parameter Θ is permitted under the integral sign." There are nontrivial cases where this does not hold (see [3]), but fortunately we will not need to be concerned with them in this book. Note, in particular, that since

$$\int_{R^n} p(v|\Theta)\, d|v| = 1 ,$$

we have that

$$\int_{R^n} \nabla_\Theta p(v|\Theta)\, d|v| = 0 ,$$

or, since

$$\nabla_\theta \log p(v|\Theta) = \frac{\nabla_\Theta p(v|\Theta)}{p(v|\Theta)}$$

(where we are not concerned with those values of v for which the denominator is zero), we can write equivalently

$$\int_{R^n} (\nabla_\Theta \log p(v|\Theta))\, p(v|\Theta)\, d|v| = 0 \qquad (3.1.4)$$

or

$$E_\Theta(\nabla_\Theta \log p(v|\Theta)) = 0$$

for all Θ in R^m. As we have seen, the bias

$$b(\Theta) = E[\hat{\Theta}] - \Theta$$

$$= \int f(v)\, p(v|\Theta)\, d|v| - \Theta$$

is a function of Θ. Since we can differentiate under the integral sign, we have that

$$\nabla_\Theta b(\Theta) = \int f(v)\, \nabla_\Theta p(v|\Theta)\, dv - I ,$$

where I is the $m \times m$ identity matrix; and we can rewrite this as

$$I + \nabla_\Theta b(\Theta) = \psi(\Theta) , \qquad (3.1.5)$$

where, using (3.1.4), we can write finally

$$\psi(\Theta) = \int_{R^n} (f(v) - \Theta)(\nabla_\Theta \log p(v|\Theta))p(v|\Theta)\, d|v| . \qquad (3.1.6)$$

The C-R inequality is basically a direct consequence of the well-known Schwarz inequality. To see this, let

$$A = f(v) - \Theta \quad ,$$

$$B = \nabla_\Theta \log p(v|\Theta) \quad .$$

Note that A is $m \times 1$ and B is $1 \times m$, so that B^* is $m \times 1$. Let Λ be any $m \times m$ matrix. Then we note that

$$((A - \Lambda B^*)(A - \Lambda B^*)^*) \geq 0$$

and hence

$$\int_{R^n} (A - \Lambda B^*)(A - \Lambda B^*)^* p(v|\Theta)\ d|v| \geq 0 \quad . \tag{3.1.7}$$

Let us "expand" (3.1.7): we have

$$E_\Theta(AA^*) - \Lambda E_\Theta(B^*A^*) - E_\Theta(AB)\Lambda^* + \Lambda E_\Theta(B^*B)\Lambda^* \geq 0 \quad . \tag{3.1.8}$$

We now assume that the $m \times m$ matrix

$$E_\Theta(B^*B) = \int_{R^n} (\nabla_\Theta \log p(v|\Theta))^*(\nabla_\Theta \log p(v|\Theta))\ p(v|\Theta)\ \ d|v|$$

is nonsingular for all Θ. Then we can choose, for each Θ,

$$\Lambda = E_\Theta(AB)\ (E_\Theta(B^*B))^{-1}$$

in (3.1.8), yielding

$$E_\Theta(AA^*) - E_\Theta(AB)(E_\Theta(B^*B))^{-1}E_\Theta(B^*A^*) \geq 0 \quad ,$$

which is readily recognized to be the same as (3.1.2).

An important question is: When can we find an estimator which actually attains the minimum second moment matrix? In other words, when does equality hold in (3.1.2)? An estimate for which equality holds in (3.1.2) is said to be "efficient." Now, equality holds in (3.1.2) if equality holds in (3.1.7), or

$$A - \Lambda B^* = 0 \quad ,$$

or $f(\cdot)$ satisfies

$$f(v) - \Theta = \Lambda(\nabla_\Theta \log p(v|\Theta))^* \qquad (3.1.9)$$

(omitting values of v for which $p(v|\Theta) = 0$). But from (3.1.9) we see that

$$E_\Theta(f(v) - \Theta) = \Lambda E_\Theta(\nabla_\Theta \log p(v|\Theta))^*$$
$$= 0 \quad ,$$

or the estimate $f(v)$ is unbiased. Moreover, the second moment matrix (which is now also the variance matrix) of the error is

$$E_\Theta((f(v) - \Theta)(f(v) - \Theta)^*)$$
$$= \Lambda E_\Theta[(\nabla_\Theta \log p(v|\Theta))^*(\nabla_\Theta \log p(v|\Theta))]\Lambda^*$$

or

$$= [E_\Theta((\nabla_\Theta \log p(v|\Theta))^* \nabla_\Theta \log p(v|\Theta))]^{-1} \quad . \qquad (3.1.10)$$

Hence for the estimate $f(v)$ to be efficient we must have

$$\nabla_\Theta \log p(v|\Theta)^* = M(\Theta)(f(v) - \Theta) \qquad (3.1.11)$$

for <u>all</u> v and Θ. The most important instance of (3.1.10) is the case where $f(v)$ is linear in v.

Finally we note an alternate formula for calculating the right side of (3.1.10) (whose inverse is the C-R bound for unbiased estimates). Let $\Theta = \text{Col.}(\theta_1, \ldots, \theta_m)$. Then

$$E[(\nabla_\Theta \log p(v|\Theta))^*(\nabla_\Theta \log p(v|\Theta))]$$

$$= -\left\{E\left(\frac{\partial^2}{\partial\theta_i \partial\theta_j} \log p(v|\Theta)\right)\right\} . \tag{3.1.12}$$

This follows from

$$0 = \frac{\partial^2}{\partial\theta_i\partial\theta_j} \int_{R^n} p(v|\Theta) \; d|v|$$

$$= \int_{R^n} \frac{\partial^2}{\partial\theta_i\partial\theta_j} p(v|\Theta) \; d|v|$$

$$= \int_{R^n} \frac{\partial}{\partial\theta_j}\left(\left(\frac{\partial}{\partial\theta_i} \log p(v|\theta)\right) p(v|\theta)\right) \; d|v|$$

$$= \int_{R^n} \left(\frac{\partial}{\partial\theta_i} \log p(v|\Theta)\right)\left(\frac{\partial}{\partial\theta_j} \log p(v|\theta)\right) p(v|\theta) \; d|v|$$

$$+ \int_{R^n} \left(\frac{\partial^2}{\partial\theta_i\partial\theta_j} \log p(v|\theta)\right) p(v|\theta) \; d|v| \; .$$

Principle of Maximum Likelihood

So far we have discussed only the measure of goodness of an estimate but have not considered the problem of finding estimates which are optimal: in other words, which minimize the error moment matrix (for all Θ). Unfortunately, there is no systematic technique for finding the optimal estimate, even if one exists. An efficient estimate cannot always be shown to exist. We can, however, single out one prescription for estimates, which has some features to recommend it -- and, in fact, the only one that is used, for all practical purposes. This is the Principle of Maximum Likelihood. The Maximum Likelihood Estimate (MLE for short) is the one that maximizes $p(v|\Theta)$ with respect to Θ, for each v. Since at a maximum (assuming the necessary differentiability properties) the gradient must be zero, the MLE will satisfy:

$$\nabla_\Theta \log p(v|\Theta) = 0 \quad .$$

We can state one desirable property of the MLE: An efficient estimate, if it exists, is a Maximum Likelihood Estimate. To see this, let us go back to (3.1.11) which characterizes the efficient estimate $f(v)$. Let us fix v in it. Then the MLE is that value of Θ which makes the left-hand side zero. Since $M(\Theta)$ is nonsingular, this means that

$$f(v) - \Theta = 0 \quad ,$$

or $f(v)$, the efficient estimate, is also the MLE.

The maximum likelihood estimate need not, in general, be unbiased; and the calculation of the corresponding error covariance can also be nontrivial. Fortunately, we are often interested only in the "asymptotic" case: where Θ is fixed in dimension, while the dimension of v grows, as we take more and more data, for instance. Thus we may talk about an estimate being "asymptotically" unbiased and "asymptotically" efficient. As a rule, the MLE has these desirable asymptotic properties, but, of course, requires proof in each particular instance. See [3, 16] for more of the statistical literature on this.

EXAMPLES

Example 1

Our first example is perhaps the oldest one of its kind. Let

$$p(v|\Theta) = \frac{1}{(\sqrt{2\pi})^n |R|^{\frac{1}{2}}} \text{Exp} -\tfrac{1}{2}[R^{-1}(v - L\Theta), (v - L\Theta)] \quad ,$$

L being $n \times m$. We see that v is Gaussian with mean $L\Theta$ and variance matrix R. We can readily verify that

$$(\nabla_\Theta \log p(v|\Theta))^* = L^* R^{-1}(v - L\Theta) \quad .$$

We assume that

$$L\Theta = 0 \quad \text{implies} \quad \Theta = 0 \quad .$$

Or equivalently

$$L^*L \text{ is nonsingular} \quad .$$

Then

$$\begin{aligned} E[(\nabla_\Theta \log p(v|\Theta))^* (\nabla_\Theta \log p(v|\Theta))] \\ &= E[L^* R^{-1}(v - L\Theta)(v - L\Theta)^* R^{-1} L] \\ &= L^* R^{-1} L \end{aligned}$$

and is nonsingular; therefore the C-R bound matrix is

$$(L^* R^{-1} L^*)^{-1} \quad .$$

The bound is thus independent of Θ. The maximum likelihood estimate is given by

$$L^*R^{-1}(v - L\hat{\Theta}) = 0 \quad ,$$

or

$$\hat{\Theta} = (L^*R^{-1}L)^{-1} L^*R^{-1}v \quad .$$

This estimate is unbiased since

$$E[\hat{\Theta}] = (L^*R^{-1}L)^{-1} L^*R^{-1}E[v] = \Theta \quad .$$

The estimate is efficient since

$$(L^*R^{-1}L)^{-1}L^*R^{-1}v - \Theta = (L^*R^{-1}L)^{-1}L^*R^{-1}[v - L\Theta]$$

$$= (L^*R^{-1}L)^{-1}(\nabla_\Theta \log p(v|\Theta))^* \quad .$$

Example 2

Our second example is one of the few cases where an efficient estimate can be explicitly calculated, and is nonlinear. It also illustrates the fact that even an efficient estimate can leave something to be desired. Thus, let

$$v = aS + N \quad ,$$

where S and N are independent Gaussians with zero means and the identity for the covariance matrix. It is desired to estimate a^2 (corresponds to signal "power"). It is immediate that

$$p(v|a^2) = \frac{1}{\sqrt{(1+a^2)^n}} \frac{1}{(\sqrt{2\pi})^n} (\mathrm{Exp} -\tfrac{1}{2}) \frac{[v,v]}{1+a^2} \quad .$$

Hence

$$\frac{\partial}{\partial a^2} \log p(v|a^2) = -\frac{n}{2}\frac{1}{1+a^2} + \frac{1}{2}\frac{[v,v]}{(1+a^2)^2} \quad .$$

The maximum likelihood estimate is such that

$$-\frac{n}{2} + \frac{1}{2}\frac{[v,v]}{1+\hat{a^2}} = 0 ,$$

or

$$\hat{a^2} = \frac{[v,v]}{n} - 1 = f_0(v) ,$$

say. This estimate is efficient since

$$\frac{[v,v]}{n} - 1 - a^2 = \frac{2(1+a^2)^2}{n}\left(-\frac{n}{2}\frac{1}{1+a^2} + \frac{1}{2}\frac{[v,v]}{(1+a^2)^2}\right) .$$

In particular, the error

$$E[(\hat{a^2} - a^2)^2] = \frac{2(1+a^2)^2}{n} .$$

However, the estimate, even if efficient, has the drawback that there is a nonzero probability that

$$\hat{a^2} < 0 ,$$

which is undesirable since we are estimating a positive quantity. We may define a new estimate which does not have this defect by taking instead

$$\frac{f_0(v) + |f_0(v)|}{2} .$$

This estimate is biased but has a smaller mean square error!

3.2. BAYESIAN THEORY OF ESTIMATION: OPTIMAL MEAN SQUARE ESTIMATES AND CONDITIONAL EXPECTATION

We next consider the "Bayesian" view, in which we assume that Θ is also random. Thus we are given the joint density

$p(\Theta, v)$. Our criterion of goodness of any estimate $f(v)$ is again the second moment matrix:

$$E[(f(v) - \Theta)(f(v) - \Theta)^*] \quad . \tag{3.2.1}$$

The optimal "mean square" estimate is the one that minimizes (3.2.1). Let

$$f_0(v) = E[\Theta | v] \quad . \tag{3.2.2}$$

We show that (3.2.2) -- the "conditional expectation of Θ given v" -- minimizes (3.2.1). For our purpose, we can express the conditional expectation as

$$E[\Theta | v] = \int_{R^m} \Theta p(\Theta | v) \; d|\Theta| \quad , \tag{3.2.3}$$

where $p(\Theta | v)$ is the conditional probability:

$$p(\Theta | v) = \frac{p(\Theta, v)}{p(v)} \quad . \tag{3.2.4}$$

The crucial property of conditional expectation we need is that for any $m \times 1$ (Borel measurable) function $h(v)$

$$E[\Theta h(v)^*] = E[E[\Theta | v] \; h(v)^*] \quad . \tag{3.2.5}$$

This can be proved easily using (3.2.3). In fact, the right side of (3.2.5) is

$$= \int_{R^n} \left(\int_{R^m} \Theta p(\Theta | v) \; d|\Theta| \right) h(v)^* p(v) \quad d|v|$$

$$= \int_{R^n}\int_{R^m} \Theta h(v)^* p(\Theta | v) \; p(v) \quad d|\Theta| \; d|v| \quad ,$$

which by virtue of (3.2.4) yields the left side of (3.2.5).

Let us now rewrite (3.2.1) as

$$E[((f_0(v)-\Theta) + f(v) - f_0(v))((f_0(v)-\Theta) + f(v) - f_0(v))^*]$$

$$= E[(f_0(v)-\Theta)(f_0(v)-\Theta)^*] + E[(f(v)-f_0(v))(f(v)-f_0(v))^*];$$

since from (3.2.5)

$$E[(f_0(v) - \Theta)(f(v) - f_0(v))^*] = 0 ,$$

taking

$$h(v) = f(v) - f_0(v) .$$

Let us note that from (3.2.5) (or directly) we can see that

$$E[f_0(v)] = E[\Theta] , \tag{3.2.6}$$

therefore

$$E[(\Theta - f_0(v))(\Theta - f_0(v))^*] \tag{3.2.7}$$

is the (minimal) error <u>covariance</u> matrix. We can "expand" (3.2.7):

$$= E[(\Theta - f_0(v))\Theta^*] \tag{3.2.8}$$

$$= E(\Theta\Theta^*) - E(f_0(v) f_0(v)^*) . \tag{3.2.9}$$

Both follow from (3.2.5). From (3.2.8) we have the inequality:

$$E(f_0(v) f_0(v)^*) \leq E(\Theta\Theta^*) . \tag{3.2.10}$$

<u>Remark</u>. The conditional expectation $E[\Theta|v]$ is definable as a Borel function of v for any two random variables Θ, v, as soon as Θ has a finite first moment, that is

$$E[|\theta|] < \infty \quad , \qquad \Theta = \text{Col.}(\theta_1, \ldots, \theta_m) \quad .$$

The definition is based on (3.2.5) rather than on (3.2.3) and does not, in particular, require that a joint density exist. See [9], for example, for more on this. For our purpose, since we are concerned mostly with Gaussians, (3.2.3) is adequate.

Let A be any $k \times m$ matrix. Then we may consider the best mean square estimate of $A\Theta$ rather than that of Θ -- some linear combinations of the components of Θ, in other words. Let $h(v)$ be any estimate. Then

$$E[(A\Theta - AE(\Theta|v))h(v)^*] \;=\; AE[(\Theta - E(\Theta|v))h(v)^*]$$

$$= \; 0 \quad ,$$

by (3.2.5). Hence as before, it follows that $AE[\Theta|v]$ is the best mean square estimate of $A\Theta$.

An important property of the Conditional Expectation we shall need in the sequel is:

$$E[\Theta|v_2] \;=\; E\{E[\Theta|v_1,v_2] \mid v_2\} \quad , \qquad (3.2.11)$$

which is easy to verify.

3.3. GAUSSIAN DISTRIBUTIONS: CONDITIONAL DENSITY; UNCONDITIONAL MAXIMUM LIKELIHOOD; MUTUAL INFORMATION

Let us now specialize to the case where Θ, v is (jointly) Gaussian. Let us see how to calculate the conditional expectation $E[\Theta|v]$ in this case. Let

$$\tilde{\Theta} = \Theta - E[\Theta] \quad ,$$

$$\tilde{v} = v - E[v] \quad ,$$

$$R_\Theta = E[\tilde{\Theta}\tilde{\Theta}^*] \quad ,$$

$$R_v = E[\tilde{v}\tilde{v}^*] \quad ,$$

$$R_{\Theta v} = E[\tilde{\Theta}\tilde{v}^*] \quad ,$$

and let Λ be the $(m+n) \times (m+n)$ compound matrix:

$$\Lambda = \begin{bmatrix} R_\Theta & R_{\Theta v} \\ R_{\Theta v}^* & R_v \end{bmatrix} .$$

Then assuming Λ to be nonsingular and writing

$$Z = \begin{vmatrix} \Theta \\ v \end{vmatrix} \quad ; \qquad \tilde{Z} = \begin{vmatrix} \tilde{\Theta} \\ \tilde{v} \end{vmatrix} \quad ,$$

where Z is now an $(n+m)$-dimensional Gaussian, we have

$$p(\Theta,v) = \frac{|\Lambda|^{-\frac{1}{2}}}{(\sqrt{2\pi})^{n+m}} \operatorname{Exp} -\tfrac{1}{2}[\Lambda^{-1}\tilde{Z}, \tilde{Z}] \quad , \tag{3.3.1}$$

where we may "partition" Λ^{-1} in the same form as we did Λ.

To calculate the conditional expectation $E[\Theta|v]$, we need not use (3.2.3) but can proceed indirectly.

<u>Theorem 3.1</u>. Let Θ, v be jointly Gaussian. Then

$$E[\Theta|v] = E[\Theta] + A(\tilde{v}) \quad , \tag{3.3.2}$$

where A satisfies (the linear equation)

$$E[\tilde{\Theta}\tilde{v}^*] = AE[\tilde{v}\tilde{v}^*] \quad . \tag{3.3.3}$$

Remark. Equation (3.3.3) is a discrete version of a more general equation known as the Wiener-Hopf equation.

Proof. We shall need to use one fact that uniquely characterizes Gaussians: that "uncorrelated Gaussians are independent." Or, more specifically: suppose X is $m \times 1$, Y is $n \times 1$ and they are jointly Gaussian; furthermore, suppose the cross correlation matrix

$$E[(X - E(X))(Y - E(Y))^*] = 0 \quad .$$

Then X and Y are independent:

$$p(X,Y) = p(X)\,p(Y) \quad . \tag{3.3.4}$$

We can see this readily, assuming that the covariance matrices of the variables

$$R_X = E[(X - E(X))(X - E(X))^*] \quad ,$$

$$R_Y = E[(Y - E(Y))(Y - E(Y))^*]$$

are nonsingular. In that case the covariance matrix of $\begin{vmatrix} X \\ Y \end{vmatrix}$ is

$$= \begin{vmatrix} R_X & 0 \\ 0 & R_Y \end{vmatrix} \quad ,$$

and its inverse is

$$\begin{vmatrix} R_X^{-1} & 0 \\ 0 & R_Y^{-1} \end{vmatrix} \quad .$$

Hence we can write, using (3.3.1):

$$p(X,Y) = \frac{\left(|R_Y|R_X|\right)^{-\frac{1}{2}}}{(\sqrt{2\pi})^{n+m}} \operatorname{Exp} -\tfrac{1}{2}\{[R_X^{-1}\tilde{X},\tilde{X}] + [R_Y^{-1}\tilde{Y},\tilde{Y}]\} \quad ,$$

which is

$$= \left(\frac{|R_X|^{-\frac{1}{2}}}{(\sqrt{2\pi})^m} \operatorname{Exp} -\tfrac{1}{2}[R_X^{-1}\tilde{X},\tilde{X}]\right)\left(\frac{|R_Y|^{-\frac{1}{2}}}{(\sqrt{2\pi})^n} \operatorname{Exp} -\tfrac{1}{2}[R_Y^{-1}\tilde{Y},\tilde{Y}]\right)$$

$$= p(X)\, p(Y) \quad ,$$

so that X and Y are actually independent. If R_X or R_Y are singular, we cannot, of course, write down the probability densities; however, if R_X is singular, we can work with a submatrix which is nonsingular, whose rank is the same as that of R_X; similarly for R_Y. The result (3.3.4) would apply to the variables corresponding to the submatrices. Since the original variables can be expressed as linear combinations of these, the independence result follows even if we cannot write down (3.3.4).

Next let us note that

$$\Theta - (E(\Theta) + A\tilde{v})$$

being linear in Θ and $\tilde{v}$, is jointly Gaussian with v, and by virtue of (3.3.3) is uncorrelated with v, since

$$E[(\Theta - E(\Theta) - A\tilde{v})\tilde{v}^*] = E[\tilde{\Theta}\tilde{v}^*] - AE[\tilde{v}\tilde{v}^*]$$

$$= 0 \quad ; \qquad (3.3.5)$$

hence $\tilde{\Theta} - A\tilde{v}$ is independent of v. Hence, in particular, for any $m \times 1$ function $h(v)$ (with finite second moment):

$$E[(\tilde{\Theta} - A\tilde{v})h(v)^*] = (E[\tilde{\Theta} - A\tilde{v}])E[h(v)]^*$$

$$= 0 \quad . \tag{3.3.6}$$

Or

$$E[(\Theta - (E[\Theta] + A(v-E[v])))h(v)^*] = 0 \quad .$$

Or, from the definition (3.2.5) of conditional expectations, our theorem follows.

Remark. The Gaussian case is thus characterized by the fact that (3.3.5) implies (3.3.6). Also, the best mean square estimate is linear in v.

Let us now calculate the corresponding error variance matrix:

$$P = E[(\Theta - E[\Theta|v])(\Theta - E[\Theta|v])^*]$$

$$= E[(\tilde{\Theta} - E[\tilde{\Theta}|v])(\tilde{\Theta} - E[\tilde{\Theta}|v])^*]$$

$$= E[(\tilde{\Theta} - E[\tilde{\Theta}|v])\tilde{\Theta}^*]$$

$$= R_\Theta - AE[\tilde{v}\tilde{\Theta}^*]$$

$$= R_\Theta - AR_{\Theta v}^* \quad . \tag{3.3.7}$$

Also specializing (3.2.9), we have

$$P = R_\Theta - E[E[\tilde{\Theta}|v]\ E[\tilde{\Theta}|v]^*] \tag{3.3.8}$$

$$= R_\Theta - AR_vA^* \quad . \tag{3.3.9}$$

Moreover, if R_v is nonsingular, we have that

$$A = R_{\Theta v}\ R_v^{-1} \tag{3.3.10}$$

so in that case

$$P = R_{\Theta} - R_{\Theta v} R_v^{-1} R_{\Theta v}^* \quad . \tag{3.3.11}$$

In particular, always:

$$P \leq R_{\Theta} \quad . \tag{3.3.12}$$

Calculating the Conditional Density $P(\Theta|v)$

It is interesting to note that (3.3.2) can be used to calculate the conditional density: $p(\Theta|v)$. We know that the mean is given by (3.3.2) and the variance P defined by (3.3.7). Let us assume that P is nonsingular: therefore R_{Θ} will be nonsingular also. Let us assume that R_v is nonsingular. If we assume that $p(\Theta|v)$ is Gaussian, then we know it must be given by

$$p(\Theta|v) = \frac{1}{(\sqrt{2\pi})^m |P|^{\frac{1}{2}}} \operatorname{Exp} -\tfrac{1}{2}[P^{-1}(\tilde{\Theta} - A\tilde{v}), \tilde{\Theta} - A\tilde{v}] \quad . \tag{3.3.13}$$

To show that this is indeed the correct answer, we have only to multiply the right side of (3.3.13) by $p(v)$ and show that the product is indeed $p(\Theta, v)$. For this purpose, let us first prove a result that is also of independent interest: viz:

$$|\Lambda| = |R_v| \; |P| \quad , \tag{3.3.14}$$

where $|\;|$ denotes determinant. To this end, we recall that we may perform "elementary operations" on a matrix while computing its determinant. Thus

$$|\Lambda| = \begin{bmatrix} R_\Theta & R_{\Theta v} \\ R^*_{\Theta v} & R_v \end{bmatrix}_{\text{Det}}$$

$$= \begin{bmatrix} R_\Theta - AR^*_{\Theta v} & R_{\Theta v} - AR_v \\ R^*_{\Theta v} & R_v \end{bmatrix}_{\text{Det}}$$

$$= \begin{bmatrix} P & 0 \\ R^*_{\Theta v} & R_v \end{bmatrix}_{\text{Det}} ,$$

and by the rules of computing the determinant of a partitioned matrix:

$$|\Lambda| = |P| \; |R_v| ,$$

as required. Hence we see that, in multiplying the right side of (3.3.13) by $p(v)$, the constant factors check. Next we need to show that

$$[P^{-1}(\tilde{\Theta} - A\tilde{v}), \tilde{\Theta} - A\tilde{v}] + [R_v^{-1}\tilde{v}, \tilde{v}] = [\Lambda^{-1}\tilde{Z}, \tilde{Z}] ,$$

where

$$\tilde{Z} = \begin{vmatrix} \tilde{\Theta} \\ \tilde{v} \end{vmatrix} .$$

In other words, we have to show that we may partition Λ^{-1} as

$$\Lambda^{-1} = \begin{bmatrix} P^{-1} & -P^{-1}A^* \\ -AP^{-1} & R_v^{-1} + A^*P^{-1}A \end{bmatrix} ,$$

or we have only to show that the product matrix

$$\begin{vmatrix} R_\Theta & R_{\Theta v} \\ R^*_{\Theta v} & R_v \end{vmatrix} \times \begin{vmatrix} P^{-1} & -P^{-1} \\ A^*P^{-1} & P_v^{-1}+A^*P^{-1}A \end{vmatrix}$$

equals the Identity (n+m) matrix. But this follows since, by the rules of multiplying partitioned matrices, the product is

$$= \begin{vmatrix} (R_\Theta - R_{\Theta v}A^*)P^{-1} & -R_\Theta P^{-1}A + R_{\Theta v}A^*P^{-1}A + R_{\Theta v}R_v^{-1} \\ (R^*_{\Theta v} - R_v A^*)P^{-1} & -R^*_{\Theta v}P^{-1}A + R_v A^* P^{-1}A + R_v R_v^{-1} \end{vmatrix}$$

and the result follows since P is self-adjoint:

$$P = R_\Theta - AR^*_{\Theta v} = R_\Theta - R_{\Theta v}A^* \quad ;$$

$$R^*_{\Theta v} - R_v A^* = 0 \quad ;$$

$$-R_\Theta P^{-1}A + R_{\Theta v}A^*P^{-1}A = (-R_\Theta + R_v A^*)P^{-1}A$$

$$= -A$$

and

$$R_{\Theta v}R_v^{-1} = A \; .$$

Remark. The formula (3.3.14) does not tell us much about P, since it involves only its determinant. If, however, Θ is scalar, then

$$P = |P| \; .$$

Thus we have

$$\text{Error Variance} = \frac{|\Lambda|}{|R_v|} \; , \qquad (3.3.14a)$$

and, of course, v can be of arbitrary dimension. Note that

$|R_v|$ in that case is the cofactor of the "1-1 position," since we can write

$$\Lambda = \begin{vmatrix} \sigma_\theta^2 & \Lambda_{1v} \\ \Lambda_{1v}^* & R_v \end{vmatrix} ,$$

where

$$\sigma_\theta^2 = E[\tilde{\Theta}\tilde{\Theta}^*] ,$$

$$\Lambda_{1v} = E[\tilde{\Theta}\tilde{v}^*] ,$$

and is $1 \times m$.

Unconditional Maximum Likelihood

We shall now indicate another way of deriving (3.3.2): by the principle of "unconditional maximum likelihood," the qualification "unconditional" indicating the slight difference from the principle of maximum likelihood we saw in Section 3.1. Thus we maximize the ("unconditional" or "joint") density function $p(v,\Theta)$, which is the basic datum of our problem under the Bayesian assumption. Or, we seek that value of Θ for which

$$\nabla_\Theta \log p(v,\Theta) = 0 \tag{3.3.2a}$$

for each v. But

$$p(v,\Theta) = p(\Theta|v)\, p(v) ,$$

and hence we only need take "the root of the gradient equation":

$$\nabla_\Theta \log p(\Theta|v) = 0 .$$

Now, we have seen that $p(\Theta|v)$ is Gaussian with mean $E[\Theta|v]$, and the Gaussian density function attains its maximum at the mean. Hence the maximum unconditional likelihood estimate (MULE) is the same as the conditional expectation. The point in showing this is that (3.3.2a) is "operationally easier" for an important class of applications we shall deal with. (Cf. Section 3.5.)

Mutual Information

Let us next calculate the mutual information $I(\Theta;v)$, that is, "information about Θ given by v," which by definition (in our context) is

$$I(\Theta;v) = E\left(\log \frac{p(\Theta,v)}{p(\Theta)p(v)}\right) .$$

We can show that

$$I(\Theta;v) = \tfrac{1}{2} \log \frac{|R_\Theta|}{|P|} . \tag{3.3.15}$$

Let us outline the steps. First

$$\log \frac{p(\Theta,v)}{p(\Theta)p(v)} = \log p(\Theta|v) - \log p(\Theta) ,$$

therefore

$$I(\Theta;v) = E[\log p(\Theta|v)] - E[\log p(\Theta)] .$$

We note that $p(\Theta|v)$ is Gaussian and so is $p(\Theta)$. Now, if X is any $m \times 1$ Gaussian with variance R_X, we can see that

$$\begin{aligned}-\log p(X) = {}& \tfrac{1}{2}\log|R_X| + \log(\sqrt{2\pi})^m \\ & + \tfrac{1}{2}[R_X^{-1}(X-E[X]),\ (X-E[X])] .\end{aligned}$$

Then

$$E[R_X^{-1}(X-E[X]), (X-E[X])] = \text{Tr. } E[R_X^{-1}(X-E[X])(X-E[X])^*]$$

$$= \text{Tr. } R_X^{-1}E[(x-E[X])(X-E[X])^*]$$

$$= \text{Tr. } R_X^{-1}R_X = m ,$$

or

$$E[-\log p(X)] = \tfrac{1}{2}\log|R_X| + \log(\sqrt{2\pi})^m + \frac{m}{2}. \quad (3.3.16)$$

Since $p(\Theta|v)$ has variance matrix P, and $p(\Theta)$ has variance matrix R_Θ, and both are $m \times m$, we obtain (3.3.15), using (3.3.16) and simplifying.

3.4. GRAM-SCHMIDT ORTHOGONALIZATION, AND COVARIANCE MATRIX FACTORIZATION

Let X be an $m \times 1$ Gaussian with zero mean, and let us use the notion

$$X = \text{Col. }(x_1, \ldots, x_m) .$$

Let R_x denote the covariance matrix of X. Then

$$R_x = \{\lambda_{ij}\} , \qquad 1 \le i, \; j \le m ,$$

where

$$\lambda_{ij} = E[x_i x_j] .$$

Let us construct a new set of Gaussian variables $\{y_i\}$, $i = 1,\ldots,n$, as follows:

$$
\begin{aligned}
y_1 &= x_1 \quad , \\
y_2 &= x_2 - E[x_2 | x_1] \; , \\
y_j &= x_j - E[x_j \mid x_1, \ldots, x_{j-1}] \quad , \\
y_m &= x_m - E[x_m \mid x_1, \ldots, x_{m-1}] \quad .
\end{aligned}
$$

Note that the $\{y_i\}$ are zero mean Gaussians, but furthermore:

$$E[y_i y_j] = 0 \quad , \qquad i \neq j \quad . \tag{3.4.1}$$

Indeed, for any i, by construction:

$$E[y_i x_j] = 0 \quad , \qquad j = 1, \ldots, i-1 \; ;$$

and since y_i is a linear combination of $x_1, \ldots, x_{j-1}, x_j$,

$$E[y_i y_j] = 0 \quad , \qquad j = 1, \ldots, i-1 \quad .$$

For $j > i$, we can exploit the fact that

$$E[y_i y_j] = E[y_j y_i]$$

to obtain (3.4.1). We say that y_i is "orthogonal" to y_j if

$$E[y_i y_j] = 0 \quad .$$

Hence the variables $\{y_i\}$ are mutually orthogonal. Equivalently, the covariance matrix of

$$Y = \text{Col.}\ [y_i, \ldots, y_m]$$

is diagonal:

$$E[YY^*] = D \quad ,$$

where

$$D = \{d_{ij}\} , \qquad d_{ij} = 0 , \quad i \neq j .$$

Note that we can express $\{y_i\}$ as

$$y_i = a_{i1}x_1 + a_{i2}x_2 + \cdots + a_{ii}x_i ,$$

where

$$a_{ii} = 1 .$$

Define the matrix L by

$$L = \{\ell_{ij}\} , \qquad 1 \leq i , \quad j \leq m ,$$

where

$$\ell_{ij} = a_{ij} , \qquad i \leq j ,$$

$$= 0 , \qquad i > j .$$

Thus defined, L is an $m \times m$ "lower-triangular" matrix, and we then obtain

$$Y = LX .$$

The determinant of a lower-triangular matrix is the product of the diagonal elements, and hence in our case

$$|L| = 1 .$$

Hence L is nonsingular, and

$$X = L^{-1}Y ,$$

where L^{-1} is also lower-triangular, with the diagonal elements also equal to unity. Furthermore, for any $p \times m$

matrix A,

$$AX = (AL^{-1})Y ,$$

so that any linear combination of the $\{x_i\}$ can also be expressed as a linear combination of the $\{y_i\}$. Moreover, we have that

$$\begin{aligned} R_x &= E[XX^*] = L^{-1} E[YY^*]L^{*-1} \\ &= L^{-1}DL^{*-1} \\ &= (L^{-1}\sqrt{D})(L^{-1}\sqrt{D})^* . \end{aligned} \qquad (3.4.2)$$

Thus we have factorized the covariance matrix R_x as

$$R_x = \mathcal{L}\mathcal{L}^* , \qquad (3.4.3)$$

where $\mathcal{L}$ is lower-triangular. Finally, let Z be an $m \times 1$ Gaussian with zero mean, and identity covariance matrix. Then

$$\mathcal{L}Z \qquad (3.4.4)$$

is Gaussian and has R_x for its covariance matrix. We note that (3.4.4) provides us with a "simulation" technique for constructing Gaussian vectors with prescribed covariance matrix, from a random number generator. Thus, let $\xi_1, \xi_2, \ldots, \xi_m$ be mutually independent random variables uniformly distributed between 0 and 1. Define

$$z_i = \Phi^{-1}(\xi_i) ,$$

where $\Phi(\cdot)$ is the cumulative Gaussian distribution

$$\Phi(z_i) = \frac{1}{\sqrt{2\pi}} \int_{-\infty}^{z_i} \exp -\frac{y^2}{2} \, dy \quad .$$

Then

$$Z = \text{Col.}\ (z_1, \ldots, z_m)$$

is Gaussian with identity covariance matrix. Using the factorization (3.4.3) and taking

$$\mathcal{L} = L^{-1}\sqrt{D} \quad ,$$

$$X = \mathcal{L}Z \quad ,$$

we can see that X is Gaussian with covariance R_x.

Remark. Even if X has nonzero mean, we may define

$$y_i = x_i - E[x_i \mid x_{i-1}, \ldots, x_1] \quad , \qquad i = 1,\ldots,m \quad ,$$

and $\{y_i\}$ would again be zero mean Gaussian, orthogonal for $i \neq j$. Also

$$y_i = \tilde{x}_i - E[\tilde{x}_i \mid \tilde{x}_{i-1}, \ldots, \tilde{x}_1] \quad ;$$

or we can write

$$Y = \text{Col.}\ \{y_i\} = L\tilde{X} = LX - LE[X]$$

and

$$X = L^{-1}Y + E[X] \quad .$$

3.5. ESTIMATION OF SIGNAL PARAMETERS IN ADDITIVE NOISE

Let us now specialize our estimation models closer to practice. We consider a communication channel with additive noise, where v_n represents the received signal at the nth sampling interval, so that we can write:

$$v_n = s_n + N_n \ , \tag{3.5.1}$$

where $\{N_n\}$ represents the channel noise and $\{s_n\}$ the transmitted signal at the nth sampling interval. The canonical problem is to estimate the signal s_n from the received waveform samples v_n. "Real time" or "on-line" operation would mean that s_n would have to be estimated from v_k, $k \leq n$. Moreover, we would need to specify the class of signals to be transmitted in some fashion. In keeping with our two points of view, we have the "sure signal" -- or "deterministic signal" -- case when we assume that signal parameters are unknown but that signals are specifiable once the parameters are specified. Or, we may have the case where the signal is a random process -- a "stochastic signal." These, of course, are not necessarily mutually exclusive points of view and further various shades in between are also often employed.

In this chapter we shall discuss only the former. We assume that the signal parameters enter linearly, and consider the following model (corresponding to processing over a fixed time interval):

$$V = \sum_1^m \theta_k S_k + N \ , \tag{3.5.2}$$

where V, S_k, N are all $p \times 1$. The $\{S_k\}$ are known to the receiver but the parameters $\{\Theta_k\}$, $k = 1,\ldots,m$, are not, and their values specify the transmitted waveform. We assume N is Gaussian with zero mean and covariance R_N. Although we have stated the problem in a Communication System setting, such a model can occur in a variety of other applications; in any event the model (3.5.2) can be considered divorced from any specific application.

To proceed with the estimation problem, it is convenient to let

$$\Theta = \text{Col. } (\Theta_1, \ldots, \Theta_m)$$

and write

$$\sum_1^m \Theta_k S_k = L\Theta \quad ,$$

where L is the $p \times m$ matrix defined by

$$\begin{aligned} L &= \{\ell_{ij}\} \quad , \\ \ell_{ij} &= i^{\text{th}} \text{ component of } S_j \qquad (3.5.3) \\ &= [S_j, e_i] \quad , \end{aligned}$$

where $\{e_i\}$ are $p \times 1$ unit basis vectors:

$$e_i = \text{Col. } [e_{i1}, e_{i2}, \ldots, e_{im}] \quad ,$$

where

$$\begin{aligned} e_{ij} = \delta_j^i &= 1 \quad , \qquad i = j \quad , \\ &= 0 \quad , \qquad i \neq j \quad . \end{aligned}$$

Assuming that Θ is an unknown parameter, we see that V is

Gaussian with mean $L\Theta$ and variance R_N. Then, $p(V|\Theta)$ denoting the density of V, we have

$$\nabla_\Theta \log p(V|\Theta) = -\tfrac{1}{2}\nabla_\Theta [R_N^{-1}(V-L\Theta), (V-L\Theta)] \quad , \qquad (3.5.4)$$

where we have tacitly assumed that R_N is nonsingular. We know that an efficient estimate exists and is given by that value of Θ for which (3.5.4) is zero.

We may also take the Bayesian view and assume that Θ is Gaussian with zero mean and variance matrix Λ. The signal is then, of course, "stochastic" in a "trivial" way. It is natural to assume that Θ is statistically independent of N ("signal and noise independent" case). We can then calculate the joint density $p(\Theta,V)$ and, in turn, calculate the conditional expectation $E[\Theta|V]$ yielding the best mean square estimate of Θ. In fact, even without explicitly calculating $p(\Theta,V)$, we know that

$$E[\Theta|V] = AV \quad ,$$

since both Θ and V have zero mean, and A is given by:

$$E[\Theta V^*] = AE[VV^*] \quad .$$

But

$$E[\Theta V^*] = E[\Theta[L\Theta+N]^*] = \Lambda L^* \quad ,$$

$$E[VV^*] = E[(L\Theta+N)(L\Theta+N)^*] = L\Lambda L^* + R_N \quad ,$$

which is singular only if <u>both</u> $L\Lambda L^*$ and R_N are singular. Assuming that at least one of them is nonsingular, we have

$$A = \Lambda L^*(L\Lambda L^* + R_N)^{-1} \quad ; \tag{3.5.5}$$

and thus our estimate is

$$\Lambda L^*(L\Lambda L^* + R_N)^{-1}v \quad . \tag{3.5.6}$$

The variance matrix of (3.5.6) is clearly

$$\Lambda L^*(L\Lambda L^* + R_N)^{-1}L\Lambda \quad ,$$

and hence the error-covariance matrix P is

$$P = \Lambda - \Lambda L^*(L\Lambda L^* + R_N)^{-1}\Lambda L \quad . \tag{3.5.7}$$

If we assume that R_N is nonsingular, we can invoke the maximum unconditional likelihood principle. Since Θ and N are independent, we have that the conditional density

$$p(V|\Theta) = p_N(V - L\Theta) \quad ,$$

$p_N(\cdot)$ being the density of N. Hence

$$\begin{aligned} \nabla_\Theta \log p(V;\Theta) &= \nabla_\Theta \log p(\Theta) + \nabla_\Theta \log p_N(V-L\Theta) \\ &= -\tfrac{1}{2}\nabla_\Theta([\Lambda^{-1}\Theta,\Theta] + [R_N^{-1}(V-L\Theta),(V-L\Theta)]) \ . \end{aligned} \tag{3.5.8}$$

The MULE makes (3.5.8) zero. But comparing with (3.5.4), we see that the difference between the MLE and MULE is the appearance of the additional term $[\Lambda^{-1}\Theta,\Theta]$ in the MULE. Or we may consider (3.5.2) as the case corresponding to

$$\Lambda = +\infty \quad ,$$

which we may interpret as "maximum ignorance"; or, in other

words, the MULE with the a priori variance equal to infinity is the same as the MLE, in this model. Let us now proceed to take the gradient in (3.5.8). We have

$$\frac{1}{2}\frac{d}{d\lambda}\left\{[\Lambda^{-1}(\Theta+\lambda h),(\Theta+\lambda h)] + [R_N^{-1}(V-L(\Theta+\lambda h)), V-L(\Theta+\lambda h)]\right\}_{\lambda=0}$$

$$= [\Lambda^{-1}\Theta, h] - [R_N^{-1}(V-L\Theta), Lh]$$

$$= [\Lambda^{-1}\Theta - L^*R_N^{-1}(V-L\Theta), h] \quad ;$$

and for the gradient to be zero this must be zero for every h in R^m; or

$$\Lambda^{-1} + L^*R_N^{-1}(V - L\Theta) = 0$$

or

$$(\Lambda^{-1} + L^*R_N^{-1}L)\Theta = L^*R_N^{-1}V \quad ,$$

or the MULE is given by

$$(\Lambda^{-1} + L^*R_N^{-1}L)^{-1} L^*R_N^{-1} \quad .$$

Hence

$$E[\Theta|V] = (\Lambda^{-1} + L^*R_N^{-1}L)^{-1}L^*R_N^{-1}V \quad , \tag{3.5.9}$$

which, under the assumption that R_N is nonsingular, must coincide with (3.5.7). Putting $\Lambda = \infty$ in (3.5.9), we get the MLE:

$$(L^*R_N^{-1}L)^{-1}L^*R_N^{-1}V \quad , \tag{3.5.10}$$

provided

$$L^*R_N^{-1}L$$

is nonsingular. Or, since R_N is assumed to be nonsingular, provided L is one-to-one: that is to say

$$L\Theta = 0$$

implies that

$$\Theta = 0 \ .$$

Or, equivalently,

$$L^*L \text{ is nonsingular} \ .$$

In terms of the $\{S_k\}$, L being nonsingular obviously means that the $\{S_k\}$ are linearly independent, a natural condition in the signal transmission context. Moreover, the estimate being efficient, the error covariance matrix equals the C-R bound which is

$$(L^* R_N^{-1} L)^{-1} \ . \tag{3.5.11}$$

Let us calculate the error covariance matrix corresponding to (3.5.9). From (3.3.7), this is

$$= \Lambda - (\Lambda^{-1} + L^* R_N^{-1} L)^{-1} L^* R_N^{-1} E(V\Theta^*)$$

$$= \Lambda - (\Lambda^{-1} + L^* R_N^{-1} L)^{-1} L^* R_N^{-1} L\Lambda$$

$$= (\Lambda^{-1} + L^* R_N^{-1} L)^{-1} ((\Lambda^{-1} + L^* R_N^{-1} L)\Lambda - L^* R_N^{-1} L\Lambda)$$

$$= (\Lambda^{-1} + L^* R_N^{-1} L)^{-1} \ . \tag{3.5.12}$$

Of course, (3.5.12) is readily seen to be smaller than (3.5.11), as it should be, since the latter corresponds to maximum ignorance. And (3.5.12) checks with (3.5.11) upon setting $\Lambda = +\infty$. If P denotes the error covariance matrix, we can write (3.5.10) and (3.5.9) as

$$\Theta = PL^* R_N^{-1} V \quad . \tag{3.5.13}$$

We note that in terms of the $\{S_k\}$

$$L^* R_N^{-1} V = \text{Col.}\,([S_1, R_N^{-1}V], \ldots, [S_m, R_N^{-1}V]) \quad .$$

In most applications we may take R_N to be a multiple of the identity:

$$R_N = dI \quad ;$$

in that case

$$P = \left(\Lambda^{-1} + \frac{L^* L}{d}\right)^{-1} \quad , \tag{3.5.14}$$

where

$$L^* L = \{[S_i, S_j]\} \quad , \qquad 1 \le i \; , \; j \le m \; .$$

<u>Remark 1</u>. If we do not care about optimality, and simply estimate Θ as

$$(L^* L)^{-1} L^* V \quad ,$$

which yields the right answer if there is no noise at all, we see that the corresponding error is

$$\Theta - (L^* L)^{-1} L^* V = (L^* L)^{-1} L^* N \quad ,$$

so that the error covariance is

$$(L^* L)^{-1} L^* R_N \, L (L^* L)^{-1} \quad ,$$

which is, of course, always

$$\ge (L^* R_N^{-1} L)^{-1} \quad .$$

This is a direct consequence of our theory although tedious

to prove directly. It is immediate, however, in the important case

$$R_N = dI .$$

Remark 2. Similarly, the proof of the equivalence of (3.5.6) and (3.5.9) in the case R_N is nonsingular, is an algebraic exercise, which we now outline for the curious. We rewrite (3.5.5) as

$$A(L\Lambda L^* + R_N) = \Lambda L^* .$$

Multiplying both sides by R_N^{-1} on the right, we have

$$AL\Lambda L^* R_N^{-1} + A = \Lambda L^* R_N^{-1} , \qquad (3.5.15)$$

or

$$A = (L + AL)L^* R_N^{-1} .$$

Hence writing

$$A = H\Lambda L^* R_N^{-1}$$

and substituting in (3.5.5), we obtain

$$H\Lambda L^* R_N^{-1} L\Lambda L^* R_N^{-1} + H\Lambda L^* R_N^{-1} = \Lambda L^* R_N^{-1} .$$

Hence it is enough if

$$H\Lambda L^* R_N^{-1} + H = I ,$$

or

$$H = (I + \Lambda L^* R_N^{-1} L)^{-1}$$

$$= (\Lambda^{-1} + L^* R_N^{-1} L)^{-1} \Lambda^{-1} ;$$

hence

$$A = (\Lambda^{-1} + L^* R_N^{-1} L)^{-1} L^* R_N^{-1} ,$$

as required.

3.6. PERFORMANCE DEGRADATION DUE TO PARAMETER UNCERTAINTY

We pause now, before treating more general estimation problems, to dwell on an important consideration in the practical utilization of our theory. We will need to deal with it more generally later, but it is instructive to see it in our present context.

As we have noted, in the Bayesian approach it is necessary to specify the a priori probability density of the unknown parameter-vector Θ. In the Gaussian case we have been specializing to, this means specifying the covariance matrix Λ. This is in general unknown, and one can argue that it is unknowable and hence must be replaced by a "guesstimate" -- say P_0. The question then is: What is a good guesstimate? Let us ponder over this briefly. Based on our theory, our estimate will then use P_0 in place of Λ:

$$\hat{\Theta} = (P_0^{-1} + L^* R_N^{-1} L)^{-1} L^* R_N^{-1} v \quad .$$

This estimate will, of course, no longer necessarily be optimal. Let us calculate the corresponding error covariance. We have

$$\hat{\Theta} - \Theta = (P_0^{-1} + L^* R_N^{-1} L)^{-1} (L^* R_N^{-1} v - (P_0^{-1} + L^* R_N^{-1} L)\Theta) \quad .$$

Let us use the notation

$$P_c = (P_0^{-1} + L^* R_N^{-1} L)^{-1} \quad ,$$

where the subscript c is supposed to indicate "calculated." Then substituting for v, we have:

$$\hat{\Theta} - \Theta = P_c(L^* R_N^{-1}(L\Theta + N) - (P_0^{-1} + L^* R_N^{-1} L)\Theta)$$

$$= P_c[L^* R_N^{-1} N - P_0^{-1}\Theta] \quad ,$$

from which is follows that the error matrix P is given by

$$P = P_c(L^* R_N^{-1} L + P_0^{-1} \Lambda P_0^{-1}) P_c \quad ,$$

which we can rewrite as

$$= P_c + P_c(P_0^{-1} \Lambda P_0^{-1} - P_0^{-1}) P_c \quad .$$

Hence the discrepancy between the actual and the calculated:

$$(P - P_c) = P_c(P_0^{-1} - P_0^{-1} \Lambda P_0^{-1}) P_c$$

$$= P_c P_0^{-1}(P_0 - \Lambda) P_0^{-1} P_c \quad . \tag{3.6.1}$$

Note that the matrix

$$D = P_0^{-1}(P_0 - \Lambda) P_0^{-1}$$

is self-adjoint and nonnegative/nonpositive definite according as $(P_0 - \Lambda)$ is. Hence we can conclude that:

True Error Variance

$\geq$ Calculated Variance if $P_0 \geq \Lambda$; (3.6.2)

True Error Variance

$\leq$ Calculated Variance if $P_0 \leq \Lambda$. (3.6.3)

The degradation in performance even if not calculable (since Λ is unknown) is

$$(\Lambda^{-1} + L^* R_N^{-1} L)^{-1} - P_c - P_c D P_c$$
$$= (\Lambda^{-1} + L^* R_N^{-1} L)^{-1} (\Lambda^{-1} - P_0^{-1}) P_c - P_c D P_c \quad .$$

The matrix

$$L^* R_N^{-1} L \tag{3.6.4}$$

deserves to be called the signal-to-noise ratio matrix. Note that the degradation in performance is "small" when the signal-to-noise ratio is "large," the latter meaning

$$L^* R_N^{-1} L \quad >> \quad \Lambda^{-1} \quad . \tag{3.6.5}$$

This is generally true and explains incidentally why at high signal-to-noise ratio anything will work.

Let us also note that we have here a dilemma that is typical in estimation theory: we may discard the overly-pessimistic "maximum-ignorance" view and go Bayesian. But the latter has the disadvantage that the necessary statistics may have to be "guesstimated." A way out of this is usually to show that as we process more and more data -- "asymptotically" -- the estimate becomes independent of the a priori probability. We shall elaborate on this in the next section.

Finally, let us examine the dependence on the uncertainty in the noise covariance: R_N. We shall only consider the case where

$$R_N = \sigma^2 I \quad .$$

Let σ_0^2 be our "guesstimate" of σ^2. Then our estimate of Θ will be

$$\hat{\Theta} = \left(P_0^{-1} + \frac{L^*L}{\sigma_0^2}\right)^{-1} \frac{L^*v}{\sigma_0^2}$$

and the corresponding error covariance P will be

$$= E[(\Theta - \hat{\Theta})(\Theta - \hat{\Theta})^*]$$

$$= P_c + P_c\left(\frac{L^*L}{\sigma_0^2}\left(\frac{\sigma^2}{\sigma_0^2} - 1\right) + \left(P_0^{-1}\Lambda P_0^{-1} - P_0^{-1}\right)\right)P_c \quad . \qquad (3.6.6)$$

Note that only the ratio $\left(\frac{\sigma^2}{\sigma_0^2}\right)$ enters; this explains why the logarithmic measure is often used:

$$10 \log \frac{\sigma^2}{\sigma_0^2}$$

in "decibels," instead of the ratio itself. Note also as before that the larger the signal-to-noise ratio

$$\frac{L^*L}{\sigma_0^2}$$

the smaller the discrepancy

$$(P - P_c) \quad .$$

Note as well the interlacing of the noise variance error and the <u>a priori</u> variance error in (3.6.6). We may also define the "normalized" discrepancy as

$$(\sqrt{P_c})^{-1} (P - P_c) (\sqrt{P_c})^{-1} \qquad (3.6.7)$$

and note that this is also smaller the higher the signal-to-noise ratio.

Chapter 4.

THE KALMAN FILTER

This is the main chapter of the book and is organized as follows. We begin in Section 4.1 with the basic theory and formulas, making a compromise in generality between too many obscuring details and too little practical application. Thus we consider only the case where the observation noise is white and is independent of the Signal, although we allow the system to be time-varying. Because of the uncertainty in the initial covariances, in practice no Kalman filter can be optimal except in the steady state -- and this is by far its most important use. Hence Section 4.2 specializes to time-invariant systems and considers steady-state behavior of the filter. Section 4.3 examines the steady-state results from the frequency-domain point of view, relating them to the more classical transfer-function approach. In Section 4.4 we study a canonical application of Kalman filtering: to System Identification. In Section 4.5 we study the "Kalman smoother": the on-line version of two-sided interpolation. In Sections 4.6 and 4.7 we study generalizations of the basic theory of Section 4.1. Thus we allow the signal and noise to be correlated in Section 4.6; and allow the observation noise to be non-white in Section 4.7. We conclude in Section 4.8 with a simple example which illustrates some of the theory and techniques discussed in the chapter.

4.1. BASIC THEORY

The estimation techniques of the previous chapter involved "batch" processing -- processing of all the data at once. We wish now to consider the "on-line" problem: where the data has to be processed sequentially, as sample-time progresses. Or, we need to design a "filter." Let us examine this more closely. We have as before the data model:

$$v_n = s_n + N_n^o , \qquad n \geq 1 , \tag{4.1.1}$$

where $\{s_n\}$ is the signal and $\{N_n^o\}$ the noise. The problem we pose now is that of estimating s_n from all the available data at sample time n. Our estimation criterion being Bayesian mean square error, to determine the optimal estimate, as we have seen, we must calculate the conditional expectation

$$\hat{s}_n = E[s_n \mid v_n, v_{n-1}, \ldots, v_1] . \tag{4.1.2}$$

We propose to do this by a (time varying in general) filter. In block diagram we have:

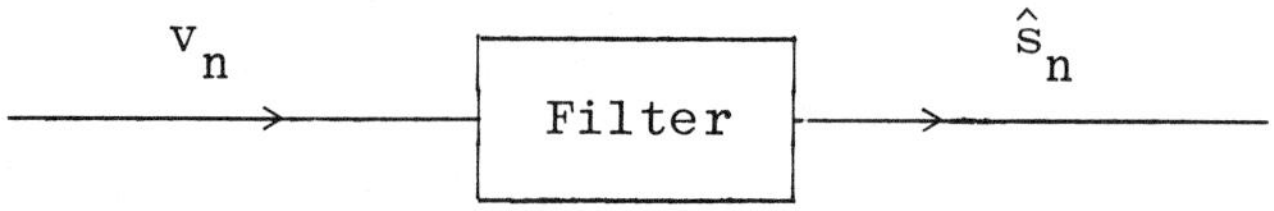

Since we will only be concerned with the Gaussian case, we know that the filter will be a "linear system" with input-output relations:

$$\hat{s}_n = \sum_1^n W_{n,k}\tilde{v}_k + E[s_n] \quad , \tag{4.1.3}$$

where

$$\tilde{v}_n = v_n - E[v_n] \quad ;$$

$$\tilde{s}_n = s_n - E[s_n] \quad ,$$

and the "system matrix" $\{W_{n,k}\}$ is determined by solving the Wiener-Hopf equation:

$$E[\tilde{s}_n v_k^*] = \sum_{j=1}^n W_{n,k} E[\tilde{v}_j \tilde{v}_k^*] \quad .$$

For each n, we see from (4.1.3) that we need to "store" all the data samples v_k, up to $k = n$. This can lead to prohibitively large data storage, and so alternate means need to be found for practical implementation.

The key is provided by having an appropriate signal-generation model for the signal $\{s_n\}$, such as we have studied in Chapter 2. Thus we now assume we have the "state space" model:

$$\left.\begin{aligned} s_n &= C_n x_n \\ x_{n+1} &= A_n x_n + U_n + N_n^s \end{aligned}\right\} , \tag{4.1.4}$$

where $\{N_n^s\}$ is a white noise sequence, $\{U_n\}$ is a known deterministic input, and of course, the matrices A_n and C_n are known. Next we assume that the "observation" noise $\{N_n^o\}$ is white also. It is then convenient to combine the white noise processes in (4.1.4) and (4.1.1), using the following notation:

$$\begin{aligned} N_n^s &= F_n N_n , \\ N_n^o &= G_n N_n , \end{aligned} \tag{4.1.5}$$

where $\{N_n\}$ is white Gaussian with unit covariance matrix, and of course:

$$\begin{aligned} E[N_n^s N_n^{s*}] &= F_n F_n^* , \\ E[N_n^o N_n^{o*}] &= G_n G_n^* , \\ E[(F_n N_n)(G_m N_m)^*] &= F_n G_m^* . \end{aligned}$$

Then we can write our signal-generation model as:

$$\left.\begin{aligned} v_n &= s_n + G_n N_n \\ s_n &= C_n x_n \\ x_{n+1} &= A_n x_n + U_n + F_n N_n \end{aligned}\right\} . \tag{4.1.4a}$$

We shall assume that

$$F_n G_m^* = 0$$

implying that signal and noise are mutually independent processes. Another important assumption throughout will be that the observation noise covariance is nonsingular:

$$G_n G_n^* \quad \text{nonsingular} \quad . \tag{4.1.6}$$

The remarkable achievement of Kalman and Bucy [8] was to show that given the signal-generation model (4.1.4a) (and the assumption that the observation noise is white) it is possible to describe the optimal linear filter also in state-space form analogous to (4.1.4). This representation is now generally referred to as the "Kalman filter," which we shall now proceed to derive.

First of all let

$$\hat{x}_n = E[x_n \mid v_n, \ldots, v_1] \quad , \qquad n \geq 1 \quad ,$$

which we refer to as the "state estimate." We assume that x_0 is Gaussian and independent of the noise sequence $\{N_n\}$, so that in particular, we may define:

$$\hat{x}_0 = E[x_0] \quad .$$

The Kalman filter is structured on two simple but essential ideas:

(i) The Innovation Sequence

We carry out a Gram-Schmidt orthogonalization of the

sequence $\{v_n\}$. Thus we define:

$$\left.\begin{aligned} \nu_n &= v_n - E[v_n \mid v_{n-1}, \dots, v_1] \\ \nu_1 &= v_1 - E[v_1] \quad \text{for} \quad n = 1 \end{aligned}\right\} . \tag{4.1.7}$$

We know of course that $\{\nu_n\}$, which is referred to as the "Innovation Sequence" (the name indicative of the fact that ν_n for each n represents the "new information" provided by the nth data sample v_n), is a white noise sequence. We can "simplify" the right side of (4.1.7) by noting that

$$\begin{aligned} v_n &= C_n x_n + G_n N_n \\ &= C_n(A_{n-1}x_{n-1} + U_{n-1} + F_{n-1}N_{n-1}) + G_n N_n . \end{aligned} \tag{4.1.8}$$

Hence we may proceed to take the conditional expectation term-by-term. We note first, however, that by the independence of signal and noise

$$E[F_{n-1}N_{n-1} \mid v_{n-1}, \dots, v_1] = 0$$

and

$$E[N_n \mid v_{n-1}, \dots, v_1] = 0 .$$

Hence

$$E[v_n \mid v_{n-1}, \dots, v_1] = C_n(A_{n-1}\hat{x}_{n-1} + U_{n-1}) ,$$

and hence we have the representation:

$$\nu_n = v_n - C_n(A_{n-1}\hat{x}_{n-1} + U_{n-1}) . \tag{4.1.9}$$

As in any Gram-Schmidt orthogonalization procedure, we know that (cf. Chapter 3)

$$E[\nu_n] = 0$$

and in fact that ν_n does not "depend" on the mean of the process $\{v_n\}$. It is convenient to express this explicitly. Let

$$\tilde{v}_n = v_n - E[v_n] \quad ,$$

$$\tilde{x}_n = x_n - E[x_n] \quad .$$

Then, as we know from Chapter 3, we can write

$$\hat{x}_n = E[x_n] + E[\tilde{x}_n \mid \tilde{v}_n, \ldots, \tilde{v}_1] \tag{4.1.7a}$$

and in particular, as a consequence,

$$\nu_n = \tilde{v}_n - E[\tilde{v}_n \mid \tilde{v}_{n-1}, \ldots, \tilde{v}_1] \quad , \tag{4.1.9a}$$

$$= \tilde{v}_n - C_n A_{n-1} \hat{\tilde{x}}_{n-1} \quad , \tag{4.1.9b}$$

where

$$\hat{\tilde{x}}_n = E[\tilde{x}_n \mid \tilde{v}_n, \ldots, \tilde{v}_1] \quad , \tag{4.1.10}$$

and we note that

$$\hat{x}_n = E[x_n] + \hat{\tilde{x}}_n \quad . \tag{4.1.10a}$$

The form (4.1.9a) shows explicitly that the innovation process ν_n remains the same whatever the deterministic input $\{U_n\}$, since the latter enters only in the definition of the mean of $\{v_n\}$. In fact, let

$$m_n = E[x_n] \quad .$$

Then, taking expectations in the state equation (4.1.4), we have:

$$m_n = A_{n-1}m_{n-1} + U_{n-1} , \qquad n \geq 1 ,$$

$$m_0 = E[x_0] , \tag{4.1.11}$$

and, of course,

$$E[v_n] = C_n m_n . \tag{4.1.12}$$

(ii) State-Innovation Sequence

Next we (Gram-Schmidt) orthogonalize the state estimate $\{\hat{x}_n\}$. Let

$$\nu_n^s = \hat{x}_n - E[\hat{x}_n \mid \hat{x}_{n-1}, \ldots, \hat{x}_1] , \tag{4.1.13}$$

$$\nu_1^s = \hat{x}_1 - m_1$$

$$= \hat{x}_1 - (A_0 m_0 + U_0) .$$

We shall refer to $\{\nu_n^s\}$ as the State-Innovation Sequence: the superscript s signifying this. As before, we may remove the means:

$$\nu_n^s = \hat{\hat{x}}_n - E[\hat{\hat{x}}_n \mid \hat{\hat{x}}_{n-1}, \hat{\hat{x}}_{n-2}, \ldots, \hat{\hat{x}}_1] .$$

To calculate the second term on the right, let us observe that

$$\hat{\hat{x}}_n - E[\hat{\hat{x}}_n \mid \hat{\tilde{v}}_{n-1}, \ldots, \hat{\tilde{v}}_1]$$

is uncorrelated with $\tilde{v}_{n-1}, \ldots, \tilde{v}_1$ and hence also with $\hat{x}_{n-1}, \ldots, \hat{x}_1$, since $\hat{\hat{x}}_{n-k}$ is a linear combination of $\tilde{v}_{n-k}, \ldots, \tilde{v}_1$. Hence it follows that

$$E[\hat{\hat{x}}_n \mid \hat{\tilde{v}}_{n-1}, \ldots, \hat{\tilde{v}}_1] = E[\hat{\hat{x}}_n \mid \hat{\hat{x}}_{n-1}, \ldots, \hat{\hat{x}}_1] .$$

But the left side can be expressed

$$E[E[\tilde{x}_n \mid \tilde{v}_n, \ldots, \tilde{v}_1] \mid \tilde{v}_{n-1}, \ldots, \tilde{v}_1]$$

and this in turn by (3.2.11)

$$= E[\tilde{x}_n \mid \tilde{v}_{n-1}, \ldots, \tilde{v}_1] \quad ,$$

which we may now calculate term-by-term, using

$$\tilde{x}_n = A_{n-1}\tilde{x}_{n-1} + F_{n-1}N_{n-1} \quad . \qquad (4.1.14)$$

This yields

$$E[\tilde{x}_n \mid \tilde{v}_{n-1}, \ldots, \tilde{v}_1] = A_{n-1}\hat{\tilde{x}}_{n-1} \quad .$$

Hence we have that

$$\nu_n^s = \hat{\tilde{x}}_n - A_{n-1}\hat{\tilde{x}}_{n-1}$$

$$= \hat{x}_n - (A_{n-1}\hat{x}_{n-1} + U_{n-1}) \quad . \qquad (4.1.15)$$

Thus the state innovation is the same whatever the input sequence, and if we wish, we may obtain it by setting U_k identically to zero.

The Kalman filter is obtained by showing that

$$\nu_n^s = K_n \nu_n \quad . \qquad (4.1.16)$$

This is based on the important property of the innovation sequence $\{\nu_n\}$ that it "contains all the information in the observation sequence $\{v_n\}$." More precisely, for each n, we can express $\tilde{v}_n$ in terms of $\nu_n, \nu_{n-1}, \ldots, \nu_1$. But this is a consequence of the Gram-Schmidt orthogonalization process that we have already seen in Chapter 3. Hence in turn, it follows that ν_n^s can be expressed linearly in terms of

$\nu_n, \nu_{n-1}, \ldots, \nu_1$. Furthermore, ν_n^s is uncorrelated with ν_{n-k}, $k \geq 1$. This is because we can express ν_n^s as:

$$\begin{aligned}
\nu_n^s &= \hat{\tilde{x}}_n - A_{n-1}\hat{\tilde{x}}_{n-1} \\
&= \hat{\tilde{x}}_n - \tilde{x}_n + \tilde{x}_n - A_{n-1}(\hat{\tilde{x}}_{n-1} - \tilde{x}_{n-1} + \tilde{x}_{n-1}) \\
&= (\hat{\tilde{x}}_n - \tilde{x}_n) - A_{n-1}(\hat{\tilde{x}}_{n-1} - \tilde{x}_{n-1}) + \tilde{x}_n - A_{n-1}\tilde{x}_{n-1} \\
&= (\hat{\tilde{x}}_n - \tilde{x}_n) - A_{n-1}(\hat{\tilde{x}}_{n-1} - \tilde{x}_{n-1}) + F_{n-1}N_{n-1} \,. \qquad (4.1.17)
\end{aligned}$$

The first two terms are uncorrelated with $\tilde{v}_{n-k}$, $k \geq 1$, by the optimality of the estimates $\hat{\tilde{x}}_n$ and $\hat{\tilde{x}}_{n-1}$. As for the third term, we have

$$E[F_{n-1}N_{n-1}\tilde{v}_{n-1}^*] = E[F_{n-1}N_{n-1}(C_{n-1}\tilde{x}_{n-1} + G_{n-1}N_{n-1})^*] \,.$$

Now the state noise $F_{n-1}N_{n-1}$ is independent of x_{n-1}; and hence the above expression is

$$\begin{aligned}
&= F_{n-1}G_{n-1}^* \\
&= 0
\end{aligned}$$

by our assumption of signal-noise independence. Hence ν_n^s is uncorrelated with $\tilde{v}_{n-k}$ for $k \geq 0$, and hence with ν_{n-k} for $k \geq 1$. Hence to prove (4.1.16) we have only to define K_n so that

$$E[\nu_n^s \nu_n^*] = K_n E[\nu_n \nu_n^*] \,. \qquad (4.1.18)$$

Using K_n so defined, we have, rewriting (4.1.16) by replacing ν_n^s, ν_n by (4.1.9) and (4.1.15), respectively:

$$\hat{x}_n - (A_{n-1}\hat{x}_{n-1} + U_{n-1}) = K_n(v_n - C_n(A_{n-1}\hat{x}_{n-1} + U_{n-1})) , \tag{4.1.19}$$

or

$$\hat{x}_n = (I - K_nC_n)A_{n-1}\hat{x}_{n-1} + (I - K_nC_n)U_{n-1} + K_nv_n , \tag{4.1.19a}$$

$$\hat{x}_0 = E[x_0] ,$$

$$\hat{s}_n = C_n\hat{x}_n .$$

The Kalman filter is defined by (4.1.19a). Note that this is a time-varying linear system with the state space same as that of the signal process. The "input" is comprised of the terms

$$(I - K_nC_n)U_{n-1} + K_nv_n$$

and thus involves the observed data sequence $\{v_n\}$ as well as the input sequence $\{U_n\}$. It is important to note that (4.1.19) can also be expressed as:

$$\hat{\hat{x}}_0 = 0 , \tag{4.1.20}$$

$$\hat{\hat{x}}_n = (I - K_nC_n)A_{n-1}\hat{\hat{x}}_{n-1} + K_n(v_n - C_nm_n) ,$$

where

$$m_{n+1} = A_nm_n + U_n ,$$

$$m_0 = E[x_0] ,$$

as is readily verified. What we have done in (4.1.20) is separate the "mean" process $\{m_n\}$. It is also important to note that

$$\hat{x}_n = A_{n-1}\hat{x}_{n-1} + U_{n-1} + K_n\nu_n , \tag{4.1.21}$$

which shows that $\{\hat{x}_n\}$ is a Gaussian Markov process just as $\{x_n\}$.

To instrument the filter system (4.1.19) we only need to calculate the gain matrix K_n. For this purpose, it is convenient to introduce a new notation for the state-estimation error covariance matrix. Thus let

$$P_n = E[e_n e_n^*] \quad , \tag{4.1.22}$$

where

$$e_n = x_n - \hat{x}_n = \tilde{x}_n - \hat{\tilde{x}}_n \quad .$$

First let us derive the difference equation satisfied by the error process $\{e_n\}$. Subtracting the difference equations for $\{x_n\}$ and $\{\hat{x}_n\}$, we obtain

$$e_n = A_n e_{n-1} + F_{n-1}N_{n-1} - K_n \nu_n \quad . \tag{4.1.23}$$

Now ν_n defined by (4.1.9) can be expressed in terms of $\{e_n\}$ as:

$$\nu_n = G_n N_n + C_n x_n - C_n(A_{n-1}\hat{x}_{n-1} + U_{n-1}) \quad ;$$

and substituting for x_n in this from (4.1.4a) we have:

$$\nu_n = G_n N_n + C_n A_{n-1} e_{n-1} + C_n F_{n-1} N_{n-1} \quad , \tag{4.1.24}$$

so that substituting this into our expression (4.1.23) for e_n we have:

$$e_n = (I - K_n C_n)A_{n-1}e_{n-1} + (I - K_n C_n)F_{n-1}N_{n-1} - K_n G_n N_n \quad . \tag{4.1.25}$$

This shows that $\{e_n\}$ is also a Gaussian Markov Process, e_{n-1} being independent of the white noise

$$(I - K_n C_n)F_{n-1}N_{n-1} - K_n G_n N_n \quad ;$$

and since each of these terms is also independent, we can readily calculate that

$$P_n = (I - K_n C_n)H_{n-1}(I - K_n C_n)^* + K_n G_n G_n^* K_n^* \quad , \tag{4.1.26}$$

where we have used the notation

$$H_n = A_n P_n A_n^* + F_n F_n^* \quad . \tag{4.1.27}$$

Let us next calculate K_n from (4.1.18) by calculating the necessary covariances. First in (4.1.24), we note that e_{n-1} is independent of $C_n F_{n-1} N_{n-1}$ because of our assumption of the independence of signal and noise; in fact,

$$\begin{aligned} E((x_{n-1} - \hat{x}_{n-1})N_{n-1}^* F_{n-1}^*) &= -E(\hat{x}_{n-1} N_{n-1}^* F_{n-1}^*) \\ &= 0 \quad , \end{aligned}$$

since $v_{n-1}, \ldots, v_1$ are all independent of $F_{n-1}N_{n-1}$. Thus every term in (4.1.24) is independent of the other two, so that

$$\begin{aligned} E[\nu_n \nu_n^*] &= G_n G_n^* + C_n(A_{n-1}P_{n-1}A_{n-1}^* + F_{n-1}F_{n-1}^*)C_n^* \\ &= G_n G_n^* + C_n H_{n-1} C_n^* \quad , \end{aligned} \tag{4.1.28}$$

which is clearly nonsingular, since $G_n G_n^*$ is. Next let us calculate the cross correlation matrix

$$E[\nu_n^s \nu_n^*] \quad .$$

For this purpose, let us first use (4.1.17). We see that the first term is uncorrelated with ν_n. For the second and third terms we exploit (4.1.24), noting that e_{n-1} is uncorrelated with $F_{n-1}N_{n-1}$ under our assumption of signal-noise independence. Hence it follows that

$$\begin{aligned} E[\nu_n^s \nu_n^*] &= (A_{n-1}P_{n-1}A_{n-1}^* + F_{n-1}F_{n-1}^*)C_n^* \\ &= H_{n-1}C_n^* \quad . \end{aligned}$$

Hence we have:

$$\begin{aligned} H_{n-1}C_n^* &= K_n E[\nu_n \nu_n^*] \\ &= K_n(G_nG_n^* + C_nH_{n-1}C_n^*) \quad . \qquad (4.1.29) \end{aligned}$$

Let us now go back to (4.1.26) and note that the right side can be rewritten:

$$P_n = (I - K_nC_n)H_{n-1} - H_{n-1}C_n^*K_n^* + K(C_nH_{n-1}C_n^* + G_nG_n^*)K_n^* \quad .$$

But substituting (4.1.29) into this, we see that

$$P_n = (I - K_nC_n)H_{n-1} \quad . \qquad (4.1.30)$$

Moreover, rewriting (4.1.29) as

$$(I - K_nC_n)H_{n-1}C_n^* = K_nG_nG_n^*$$

and using (4.1.30), we can express K_n in terms of P_n as:

$$P_nC_n^* = K_nG_nG_n^* \quad .$$

Or finally we have

$$K_n = P_n C_n^*(G_n G_n^*)^{-1} \tag{4.1.31}$$

and

$$P_n = (I - P_n C_n^*(G_n G_n^*)^{-1} C_n) H_{n-1} \quad . \tag{4.1.32}$$

Collecting terms containing P_n, we have

$$P_n(I + C_n^*(G_n G_n^*)^{-1} C_n H_{n-1}) = H_{n-1} \quad . \tag{4.1.33}$$

The matrix in parentheses (see Problem 4.1.1, if necessary) is nonsingular and hence

$$P_n = H_{n-1}(I + C_n^*(G_n G_n^*)^{-1} C_n H_{n-1})^{-1} \quad , \tag{4.1.34}$$

(and taking adjoints)

$$P_n = (I + H_{n-1} C_n^*(G_n G_n^*)^{-1} C_n)^{-1} H_{n-1} \quad . \tag{4.1.35}$$

We have thus expressed P_n in terms of P_{n-1}. To determine P_n we have only to add

$$\begin{aligned} P_0 &= E[(x_0 - \hat{x}_0)(x_0 - \hat{x}_0)^*] \\ &= E[\tilde{x}_0 \tilde{x}_0^*] \\ &= \Lambda \quad , \end{aligned} \tag{4.1.36}$$

say. This completes our derivation of the Kalman filter, which we may now express as:

$$\hat{x}_n = A_{n-1}\hat{x}_{n-1} + U_{n-1} + P_n C_n^*(G_n G_n^*)^{-1}(v_n - (A_{n-1}\hat{x}_{n-1} + U_{n-1})), \tag{4.1.37}$$

$$\hat{x}_0 = E[x_0] \quad ,$$

where P_n is given by (4.1.35) (which is our "error propagation" equation), with P_0 defined by (4.1.36).

One-Step Predictor

It is often useful to express the Kalman filter formula (4.1.34) in a different way in terms of the "one-step predictor." Thus let

$$\bar{x}_{n+1} = E[x_{n+1} \mid v_1, \ldots, v_n] \quad .$$

Note that $\bar{x}_{n+1}$ is our prediction "one step ahead" of the state at $(n+1)$ given the data up to n. We may express both ν_n and ν_n^s in terms of $\bar{x}_n$, since

$$\begin{aligned} \nu_n &= v_n - E[v_n \mid v_{n-1}, \ldots, v_1] \\ &= v_n - C_n \bar{x}_n \end{aligned}$$

and

$$\begin{aligned} \nu_n^s &= \hat{x}_n - E[\hat{x}_n \mid \hat{x}_{n-1}, \ldots, \hat{x}_1] \\ &= \hat{x}_n - E[x_n \mid v_{n-1}, \ldots, v_1] \\ &= \hat{x}_n - \bar{x}_n \quad . \end{aligned}$$

Also using (4.1.4), we have:

$$\bar{x}_{n+1} = A_n \hat{x}_n + U_n \quad . \tag{4.1.38}$$

The corresponding error is

$$x_{n+1} - \bar{x}_{n+1} = A_n(x_n - \hat{x}_n) + F_n N_n \quad ;$$

and hence the one-step-predictor error-covariance:

$$E[(x_{n+1} - \bar{x}_{n+1})(x_{n+1} - \bar{x}_{n+1})^*] = A_n P_n A_n^* + F_n F_n^* = H_n$$

yielding, in particular, an "interpretation" of H_n. Moreover, we can rewrite the Kalman filter formula (4.1.37) in terms of the quantities involving one-step prediction, using (4.1.38) and (4.1.16), in the form:

$$(\hat{x}_n - x_n) = K_n(v_n - C_n \bar{x}_n) \quad .$$

Hence we have

$$\bar{x}_n = A_{n-1}(I - K_{n-1}C_{n-1})\bar{x}_{n-1} + A_{n-1}K_{n-1}v_{n-1} \quad ,$$

$$\hat{x}_n = (I - K_n C_n)\bar{x}_n + K_n v_n \quad ,$$

$$\bar{x}_1 = U_0 \; , \qquad \hat{x}_0 = 0 = \bar{x}_0 \quad , \tag{4.1.39}$$

$$K_n = H_{n-1}C_n^*(C_n H_{n-1} C_n^* + G_n G_n^*)^{-1} \quad .$$

Signal Estimate Error Covariance

The signal estimation error covariance can be expressed in terms of P_n as:

$$E[(s_n - \hat{s}_n)(s_n - \hat{s}_n)^*] = C_n P_n C_n^* \quad .$$

Let us now derive the equations that $C_n P_n C_n^*$ must satisfy. From (4.1.33), multiplying on the left by C_n and on the right by C_n^*, we have

$$\begin{aligned} C_n H_{n-1} C_n^* &= C_n P_n C_n^* + C_n P_n C_n^* (G_n G_n^*)^{-1} C_n H_{n-1} C_n^* \\ &= (C_n P_n C_n^*)(I + (G_n G_n^*)^{-1} C_n H_{n-1} C_n^*) \quad , \end{aligned} \tag{4.1.40}$$

or

$$C_n P_n C_n^* = C_n H_{n-1} C_n^* (I + (G_n G_n^*)^{-1} C_n H_{n-1} C_n^*)^{-1}$$

$$= (I + C_n H_{n-1} C_n^* (G_n G_n^*)^{-1})^{-1} C_n H_{n-1} C_n^* \quad . \qquad (4.1.41)$$

From (4.1.40) we can also write

$$C_n P_n C_n^* = C_n H_{n-1} C_n^* (I - (G_n G_n^*)^{-1} C_n P_n C_n^*) \quad . \qquad (4.1.42)$$

Fit Error

We may also consider at this point what is known as the "fit error." This is an error that we can observe: the error between the data sequence $\{v_n\}$ and our "best fit" to the data $\{\hat{s}_n\}$:

$$z_n = v_n - \hat{s}_n \quad .$$

It is easy to see that this is a white noise sequence. To obtain the fit error covariance, we note that

$$G_n N_n = v_n - s_n$$

$$= v_n - \hat{s}_n + s_n - \hat{s}_n$$

$$= z_n + C_n e_n \quad .$$

Since $(s_n - \hat{s}_n)$, by virtue of the optimality of $\hat{s}_n$ is uncorrelated with v_{n-k}, $k \geq 0$, it follows that z_n and e_n are independent, and hence, in particular,

$$G_n G_n^* = E[z_n z_n^*] + C_n P_n C_n^* \quad . \qquad (4.1.43)$$

It follows that

$$E[z_n z_n^*] = G_n G_n^* - C_n P_n C_n^* .$$

In particular:

$$E[z_n z_n^*] \le G_n G_n^*$$

$$= G_n G_n^* \quad \text{iff.} \quad \hat{s}_n = s_n .$$

From (4.1.44) we see that a fit-error variance much smaller than the noise variance indicates poor filter performance. Let us show that the fit error variance is nonsingular. For this purpose let us express z_n in terms of ν_n. We have

$$z_n = v_n - C_n \hat{x}_n$$

$$= v_n - C_n (A_{n-1}\hat{x}_{n-1} + B_{n-1}U_{n-1} + P_n C_n^* (G_n G_n^*)^{-1} \nu_n)$$

$$= (I - C_n P_n C_n^* (G_n G_n^*)^{-1}) \nu_n . \qquad (4.1.45)$$

From which we also have that

$$E[z_n z_n^*] = (I - C_n P_n C_n^* (G_n G_n^*)^{-1}) E[\nu_n \nu_n^*] (I - (G_n G_n^*)^{-1} C_n P_n C_n^*) .$$

(4.1.46)

Suppose

$$(I - (G_n G_n^*)^{-1} C_n P_n C_n^*) x = 0 , \quad x \neq 0 . \qquad (4.1.47)$$

Then by (4.1.42) we would have that

$$C_n P_n C_n^* x = 0$$

and hence from (4.1.47) that

$$x = 0 ,$$

leading to a contradiction. Hence

$$(I - (G_n G_n^*)^{-1} C_n P_n C_n^*)$$

and

$$(I - C_n P_n C_n^* (G_n G_n^*)^{-1})$$

are nonsingular. And hence

$$\begin{aligned} E[z_n z_n^*] &= G_n G_n^* - C_n P_n C_n^* \\ &= (I - C_n P_n C_n^* (G_n G_n^*)^{-1})(G_n G_n^*) \end{aligned}$$

are nonsingular. Now

$$\begin{aligned} &(I + C_n H_{n-1} C_n^* (G_n G_n^*)^{-1})(I - C_n P_n C_n^* (G_n G_n^*)^{-1}) \\ &\quad = I + C_n H_n C_n^* (G_n G_n^*)^{-1} - C_n P_n C_n^* (G_n G_n^*)^{-1} \\ &\qquad - (C_n H_n C_n^* (G_n G_n^*)^{-1} C_n P_n C_n^* (G_n G_n^*) \\ &\quad = I \ , \end{aligned}$$

since

$$C_n H_{n-1} C_n^* - C_n P_n C_n^* - C_n H_{n-1} C_n^* (G_n G_n)^{-1} C_n P_n C_n^* = 0$$

by (4.1.40). Hence we also obtain

$$(I + C_n H_{n-1} C_n^* (G_n G_n^*)^{-1}) = (I - C_n P_n C_n^* (G_n G_n^*)^{-1})^{-1} \ . \quad (4.1.48)$$

We can also reformulate the Kalman filter equations (4.1.21) in terms of the fit error rather than the innovation. Thus we have:

$$\hat{x}_n = A_{n-1} \hat{x}_{n-1} + B_{n-1} U_{n-1} + J_n z_n \ , \quad (4.1.49)$$

$$z_n = v_n - C_n \hat{x}_n \ ,$$

$$J_n = P_n C_n^*(G_n G_n^*)^{-1}(I - C_n P_n C_n^*(G_n G_n^*)^{-1})^{-1}$$

$$= P_n C_n^*(G_n G_n^*)^{-1}(I + C_n H_{n-1} C_n^*(G_n G_n^*)^{-1})^{-1} \quad . \qquad (4.1.50)$$

Remark. Before we leave this section let us make an important observation. For the filter to be optimal we need to calculate P_n by formula (4.1.32) which involves

$$P_0 = \Lambda \ .$$

But this (starting) covariance is unknown, and perhaps even unknowable. Hence no Kalman filter in practice is optimal, except in the case where the signal-generator model is time invariant; then we can show that under certain conditions we can make our filter asymptotically optimal, whatever the initial guesstimate for Λ.

★ PROBLEMS ★

Problem 4.1.1

Let L, M be self-adjoint nonnegative definite matrices. Show that $(I + LM)$ is nonsingular.

Hint:
$$(I+LM)x = 0 \;\Rightarrow\; [(I+LM)x,\ Mx] = 0$$
$$\Rightarrow\; Mx = 0 \;\Rightarrow\; x = 0 \ .$$

Problem 4.1.2

Consider the case where $\{v_n\}$ is one-dimensional. Show that if the fit error is zero, then the signal-estimation error is the maximum possible. What happens to the Kalman filter in this case?

Problem 4.1.3

Show that

$$z_n = \nu_n - C_n \nu_n^s .$$

Problem 4.1.4

In the class of estimates of s_n of the form $\alpha_n v_n$, find the optimal α_n that minimizes the error covariance $E[(s_n - \alpha_n v_n)(s_n - \alpha_n v_n)^*]$. Denoting the minimal error covariance by T_n, show that

$$C_n P_n C_n^* \leq T_n \leq G_n G_n^* .$$

Show that the optimal α_n is given by:

$$C_n R_n C_n^* (I + C_n R_n C_n^*)^{-1} , \qquad R_n = E[x_n x_n^*] .$$

Problem 4.1.5

Consider the case where there is no "state noise": $F_n \equiv 0$. Show that in that case, for zero input $(U_n \equiv 0)$ and $E[x_0] = 0$:

$$\hat{s}_n = C_n \Phi_n \left(\sum_1^n \Phi_i^* C_i^* (G_i G_i^*)^{-1} C_i \Phi_i + \Lambda^{-1} \right)^{-1} \left(\sum_1^n \Phi_i^* C_i^* (G_i G_i^*)^{-1} v_i \right) ,$$

where

$$\Phi_i = A_{i-1} A_{i-2} \cdots A_0 ,$$

$$\Phi_0 = \text{Identity} .$$

Show that

$$P_n = \Phi_n \left(\sum_1^n \Phi_i^* C_i^* (G_i G_i^*)^{-1} \Phi_i C_i + \Lambda^{-1} \right)^{-1} \Phi_n^*$$

satisfies (4.1.33) (yielding a "closed form" solution to (4.1.32)).

Hint:

$$\hat{x}_n = \Phi_n E[x_0 \mid v_n, \ldots, v_1] ,$$

and use the "batch" formulas of Chapter 3.

Problem 4.1.6

Find the optimal "zero memory"-state estimator, i.e., find $E[x_n | v_n]$ and the corresponding error matrix.

Problem 4.1.7

Let z_n denote the fit error and let for each n:

$$z = z_1, z_2, \ldots, z_n ; \qquad \tilde{v} = \tilde{v}_1, \tilde{v}_2, \ldots, \tilde{v}_n ;$$

$$\tilde{v}_n = v_n - C_n x_n .$$

Let R denote the covariance matrix of v:

$$R = E[\tilde{v}\tilde{v}^*] .$$

Show that

$$|R| = \prod_{k=1}^{n} |G_k G_k^*| \; |(I - (G_k G_k^*)^{-1} C_k P_k C_k^*)^{-1}| ,$$

where $|\;|$ denotes determinant. Similarly, writing

$$\nu = \nu_1, \nu_2, \ldots, \nu_n ,$$

show that the determinant of the covariance matrix of ν:

$$|E[\nu\nu^*]| = \prod_{k=1}^{n} |(G_k G_k^* + C_k H_{k-1} C_k^*)| = |R| .$$

Hint: Use

$$z_n = (I - C_n P_n C_n^* (G_n G_n^*)^{-1}) \tilde{v}_n$$

+ terms containing $\tilde{v}_{n-1}, \ldots, \tilde{v}_1$,

$$z = L\tilde{v} \quad ,$$

where L is "block" lower-triangular.

Problem 4.1.8

Show that

$$P_{n+1} \leq A_n P_n A_n^* + F_n F_n^* \quad .$$

Problem 4.1.9

This problem illustrates the dependence on the guesstimate for the initial covariance. Assume zero input. Define the suboptimal filter

$$\hat{x}_n^a = (I - P_n^a C_n^* C_n) A_{n-1} \hat{x}_{n-1}^a + P_n^a C_n^* v_n \quad , \qquad \hat{x}_n^a = 0 \quad ,$$

$$P_n^a = (I + H_{n-1}^a C_n^* C_n)^{-1} H_{n-1}^a \quad ,$$

$$P_0^a = \Lambda \, ,$$

$$H_n^a = A_n P_n^a A_n^* + F_n F_n^* \quad .$$

Let T_n denote the corresponding error-covariance matrix:

$$T_n = E[(x_n - \hat{x}_n^a)(x_n - \hat{x}_n^a)^*] \quad .$$

Show that

$$P_n^a - T_n = (I - P_n^a C_n^* C_n) A_{n-1} (P_{n-1}^a - T_{n-1}) A_{n-1}^* (I - C_n^* C_n P_n^a) \quad .$$

Hence in particular

$$T_n \leq P_n^a$$

if

$$\Lambda = P_0^a > P_0 = E[x_0 x_0^*] \quad .$$

Problem 4.1.10

Let

$$\hat{N}_n = E[N_n \mid v_1, \ldots, v_n] .$$

Is $\{\hat{N}_n\}$ white noise? What is the covariance? Calculate $E[\hat{N}_n N_m^*]$. Are $\{F_n \hat{N}_n\}$ and $\{G_n \hat{N}_n\}$ independent?

Problem 4.1.11

Alternate Definition of Fit Error: We may define another "fit error," using now the one-step predictor, as:

$$v_n - \bar{s}_n ,$$

where

$$\bar{s}_n = C_n \bar{x}_n .$$

Show that the variance of this fit error is equal to the variance of the innovation and is always larger than the noise variance. When does it attain its minimum?

4.2. KALMAN FILTER: STEADY STATE THEORY

By far the most important for us is the case where the signal is (asymptotically) stationary. Thus we need to condider the asymptotics of the case where the signal-generation model (4.1.4) is time-invariant. Let us restate this problem for the time-invariant system:

Since the signal-generation system is time-invariant, the concepts of Observability and Controllability can tell us much about its structure. Let us first examine what role these concepts play in analyzing filter performance.

<u>Theorem 4.2.1</u>. Suppose $(A \sim F)$ is controllable. Then if P_n is nonsingular, so is P_{n+k} defined by (4.2.2) for every integer $k \geq 0$.

<u>Proof</u>. Let P_n be nonsingular. Then H_n must also be nonsingular. For suppose for some nonzero x we have:

$$H_n x = AP_n A^* x + FF^* x = 0 .$$

Then

$$[P_n A^* x, A^* x] + [F^* x, F^* x] = 0 ,$$

hence

$$F^* x = 0 \quad \text{and} \quad P_n A^* x = 0 .$$

Since P_n is nonsingular, we must have

$$A^* x = 0 ,$$

hence

$$F^* A^{*k} x = 0 , \qquad k \geq 1 ; \quad x \neq 0 .$$

This violates controllability. Hence H_n is nonsingular. But

$$P_{n+1} = (I + H_n C^* C)^{-1} H_n$$

and hence is nonsingular. Hence by induction P_{n+k} is nonsingular for every positive integer k.

Let us next examine the role played by Observability.

$$v_n = s_n + GN_n ,$$

$$s_n = Cx_n ,$$

$$x_{n+1} = Ax_n + u_n + FN_n ,$$

$$FG^* = 0 . \qquad (4.2.1)$$

Then our filter formulas are:

$$\hat{x}_n = (I - P_n C^*(GG^*)^{-1}C)A\hat{x}_{n-1} + P_n C^*(GG^*)^{-1}v_n$$

$$= \bar{x}_n + H_n(CH_{n-1}C^* + GG^*)^{-1}(v_n - C\bar{x}_n) ,$$

$$P_{n+1} = (I + H_n C^*(GG^*)^{-1}C)^{-1}H_n ,$$

$$H_n = AP_nA^* + FF^* ,$$

$$P_0 = E[x_0x_0^*] . \qquad (4.2.2)$$

In order to save space, we shall from now on take

$$GG^* = \text{Identity } I .$$

Otherwise, we only have to replace

$$C \text{ by } \left(\sqrt{GG^*}\right)^{-1} C ;$$

$$v_n \text{ by } \left(\sqrt{GG^*}\right)^{-1} v_n .$$

We shall also set the input u_n to be zero, since our main interest is in the error covariance, which is the same regardless of the input.

Theorem 4.2.2. Suppose x is unobservable:

$$CA^n x = 0 \qquad \text{for every } n \geq 0 ,$$

and further:

$$\sup_n \|A^n x\| = +\infty . \tag{4.2.3}$$

Then we can choose P_0 so that

$$\lim_n \text{Tr. } P_n = +\infty . \tag{4.2.4}$$

Proof. We may without loss of generality assume that

$$\|x\| = 1 .$$

Let us choose P_0 so that

$$P_0 x = \lambda x , \qquad \lambda > 0 .$$

Let

$$x_0 = [x_0, x]x + z .$$

Now

$$x_{n+1} = Ax_n + FN_n$$

has the solution:

$$x_n = A^n x_0 + \sum_0^{n-1} A^{n-k-1} FN_k .$$

Hence

$$v_n = Cx_n + GN_n$$

$$= CA^n z + C\sum_0^{n-1} A^{n-k-1} FN_k + GN_n \tag{4.2.5}$$

since

$$CA^n x = 0 .$$

From (4.2.5) it follows that

$$E[[x_0, x]v_n^*] = 0$$

since

$$E[[x_0, x]z^* A^{*n} C^*] = E[[x_0, x]z^*]A^{*n}C^*$$

and

$$\begin{aligned} E[[x_0, x]z^*] &= E[[x_0, x](x_0 - [x_0, x]x)^*] \\ &= E[x^* x_0(x_0 - [x_0, x]x)^*] \\ &= x^* E[x_0 x_0^*] - x^* E[x_0 x_0^*]xx^* \\ &= \lambda x^* - \lambda x^* xx^* \\ &= 0 \ . \end{aligned}$$

Hence $\{v_n\}$ is independent of $[x_0, x]$ and hence so is x_n. Hence

$$x_n - \hat{x}_n = [x_0, x]A^n x + A^n z + \sum_0^{n-1} A^{n-k} FN_k - \hat{x}_n \ ,$$

where the second, third and fourth terms are uncorrelated with the first. Hence

$$\begin{aligned} \text{Tr. } P_n &= \text{Tr. } E[[x_n - \hat{x}_n][x_n - \hat{x}_n]^*] \\ & \quad E[[x_0, x]^2][A^n x, \ A^n x] \ . \end{aligned}$$

Since

$$E[[x_0, x]^2] = [P_0 x, \ x] = \lambda \ ,$$

we have that

$$\text{Tr. } P_n \geq \|A^n x\|^2 \lambda$$

or

$$\sup_n \text{Tr. } P_n = +\infty .$$

<u>Remark</u>. Note that even though

$$\text{Tr. } P_n = +\infty ,$$

we do have that (cf. Exercise 4.1.4)

$$CP_nC^* \leq CR_nC^*(I + CR_nC^*)^{-1} \leq I .$$

As we have already noted, it is impossible to construct (in general) an (optimal) Kalman filter, because to do so we need P_n, and P_n depends on P_0 which is unknown and unknowable. However, in the time-invariant case (4.2.1), we can show that it is possible to achieve "asymptotic" optimality. Thus under certain conditions (indicated below) we can show that P_n will converge to a limit matrix P as $n \to \infty$, independent of the initial matrix P_0, and further, if we define:

$$\hat{x}_n^a = (I - PC^*C)A\hat{x}_{n-1}^a + PC^*v_n ,$$

then the corresponding error covariance

$$E[(x_n - x_n^a)(x_n - x_n^a)^*] \to P , \quad \text{as } n \to \infty.$$

For this purpose, let us define the function $\Phi(\cdot)$ on self-adjoint nonnegative definite matrices P by

$$\Phi(P) = (I + H(P)C^*C)^{-1}H(P) , \qquad (4.2.6)$$

where $H(P)$ is defined by

$$H(P) = APA^* + FF^* .$$

Then (4.2.2) becomes:

$$P_{n+1} = \Phi(P_n) .$$

Suppose P_n converges to P_∞. Then we would have

$$P_\infty = \Phi(P_\infty) .$$

The equation

$$P = \Phi(P) \tag{4.2.7}$$

is called the Algebraic (or Steady State) Riccati Equation (SSRE); and we see that P_∞, if it exists, satisfies this equation.

Let us recall next some definitions from state space theory. The class of (C-A) unobservable states is defined as:

all x such that $CA^k x = 0$ for every $k \geq 0$.

The class of A-stable states is defined as:

all x such that $\|A^k x\| \to 0$ as $k \to \infty$.

The class of stable states is a linear subspace. Let $\mathcal{P}_s$ denote the corresponding projection. The class of ($C \sim A$) unobservable states is also a linear subspace. Let $\mathcal{P}_u$ denote the corresponding projection and $\mathcal{P}_r$ that of the orthogonal complement. Our basic result on the steady state Kalman filter is:

Theorem 4.2.3. Suppose all $(C \sim A)$ unobservable states are A-stable and all $(F^* \sim A^*)$ unobservable states are A^*-stable. Then the steady state Riccati equation has a unique self-adjoint nonnegative definite solution. Denote it P_∞:

$$\Phi(P_\infty) = P_\infty \quad .$$

Moreover,

$$(I - P_\infty C^* C)A$$

is stable, and the filter

$$\hat{x}_n^a = A\hat{x}_{n-1}^a + P_\infty C^* [v_n - CA\hat{x}_{n-1}^a] \; ; \qquad \hat{x}_0^a = 0$$

is asymptotically optimal.

Proof. Let us state the individual results we need as lemmas, since they would be of independent interest.

Lemma 4.2.1. Suppose P and Q are self-adjoint and nonnegative definite. Then

$$\Phi(P+Q) \geq \Phi(P) \quad .$$

Proof. Let

$$H(P) = APA^* + FF^*$$

mapping self-adjoint nonnegative matrices into the same class. Let P, Q be self-adjoint and nonnegative definite and let $\lambda > 0$. Then we can express $\Phi(P+Q) - \Phi(P)$ as:

$$\Phi(P+Q) - \Phi(P) = \int_0^1 \frac{d}{d\lambda} \Phi(P+\lambda Q) \, d\lambda \quad . \tag{4.2.8}$$

Now:

$$\frac{d}{d\lambda}\,\Phi(P+\lambda Q) = \frac{d}{d\lambda}\,\Phi(I + H(P+\lambda Q)C^*C)^{-1}H(P+\lambda Q)$$

$$= (I + H(P+\lambda Q)C^*C)^{-1}AQA^*$$

$$- (I + H(P+\lambda Q)C^*C)^{-1}AQA^*C^*C(I + H(P+\lambda Q)C^*C)^{-1}H(P+\lambda Q)$$

$$= (I + H(P+\lambda Q)C^*C)^{-1}AQA^*(I - C^*C\Phi(P+\lambda Q)) \quad .$$

For any self-adjoint nonnegative definite matrix M:

$$I - \Phi(M)C^*C = I - (I + H(M)C^*C)^{-1}H(M)C^*C$$

$$= (I + H(M)C^*C)^{-1}[I + H(M)C^*C - H(M)C^*C]$$

$$= (I + H(M)C^*C)^{-1} \quad .$$

Hence

$$\frac{d}{d\lambda}\Phi(P+\lambda Q) = (I - \Phi(P+\lambda Q)C^*C)AQA(I - C^*C\Phi(P+\lambda Q))$$

$$\geq 0 \quad .$$

From (4.2.8) it follows that

$$\Phi(P+Q) - \Phi(P) \geq 0 \qquad \text{if} \quad Q \geq 0 \quad .$$

Hence, in particular:

$$\Phi(P_{n+1}) \geq \Phi(P_n) \qquad \text{if} \quad P_{n+1} \geq P_n \quad .$$

Lemma 4.2.2. Suppose $(C \sim A)$ unobservable states are A-stable. Then for any choice of P_0

$$P_n \leq M \qquad \text{for all} \quad n \quad , \tag{4.2.9}$$

where M is a self-adjoint nonnegative definite matrix, and may depend on P_0.

Proof. First let us assume that we have $(C \sim A)$ observability. Assume the matrix A is $n \times n$. Define

$$R = \sum_0^{n-1} A^{*k} C^* C A^k .$$

Then R must be nonsingular, as we have seen in Chapter 1. For $m > n$, define the suboptimal estimate:

$$\hat{x}_m^s = A^n R^{-1} \sum_0^{n-1} A^{*k} C^k v_{m+k-n} .$$

What we have constructed thereby is a filter with a memory of n, but not optimal. We show that the corresponding error covariance is bounded. Exploiting the time invariance of the system (4.2.1), we have:

$$x_{m+k-n} = A^k x_{m-n} + \sum_0^{k-1} A^j F N_{m-n-j+k-1} , \qquad m \geq n .$$

Let

$$\zeta_{m-n+k} = \sum_0^{k-1} A^j F N_{n-m+k-1+j} , \qquad m \geq n .$$

Then we can express $\hat{x}_m^s$ as

$$\hat{x}_m^s = A^n R^{-1} \sum_0^{n-1} A^{*k} C^* C A^k x_{m-n}$$

$$+ \quad A^n R^{-1} \sum_0^{n-1} A^{*k} C^* (C\zeta_{n-m+k} + GN_{m-n+k})$$

$$= A^n x_{m-n} + A^n R^{-1} \sum_0^{n-1} A^{*k} C^* (C\zeta_{m-n+k} + GN_{m-n+k}) \ ;$$

hence the error in the estimate:

$$\hat{x}_m^s - x_m = A^n R^{-1} \sum_0^{n-1} A^{*k} C^* (C\zeta_{m-n+k} + GN_{m-n+k}) - \zeta_m \ .$$

It is readily verified that the covariance of the random vector on the right side is independent of m. Hence

$$E[(\hat{x}_m^s - x_m)(\hat{x}_m^s - x_m)^*] = J \ ,$$

where J is a fixed matrix. Since P_m is the minimal error matrix we have

$$P_m \leq J \ , \qquad m \geq n \ ;$$

hence, taking

$$M = \max (J, \max_{k \leq n} P_k) \ ,$$

the lemma follows.

Suppose next that A is stable: all eigenvalues are strictly less than one in magnitude. In that case the conditions of our Lemma are satisfied, and we can prove (4.2.9) very simply. We have

$$P_n = E[x_n x_n^*] - E[\hat{x}_n \hat{x}_n^*]$$

$$\leq E[x_n x_n^*] \ ,$$

and we know from Chapter 2 that

$$E[x_n x_n^*] = A^n P_0 A^{*n} + \sum_0^{n-1} A^k F F^* A^{*k} .$$

Now since A is stable, the first term goes to zero as $n \to \infty$ (regardless of the choice of P_0) and the second term converges to

$$\sum_0^{\infty} A^k F F^* A^{*k} .$$

Hence (4.2.9) is immediate.

In other words, if $(C \sim A)$ is observable or A is stable, the Lemma holds. Let us now consider the more general situation where A may be unstable, but all $(C \sim A)$ unobservable states are stable.

We begin by noting that

$$Cx_n = C\mathcal{P}_r x_n$$

and

$$\mathcal{P}_r x_n = \mathcal{P}_r A x_{n-1} + \mathcal{P}_r F N_{n-1} .$$

It is readily verified that A maps the subspace of $(C \sim A)$ unobservable states into itself:

$$A\mathcal{P}_u x = \mathcal{P}_u A \mathcal{P}_u x$$

and hence also

$$\mathcal{P}_r A x = \mathcal{P}_r A(\mathcal{P}_r x + \mathcal{P}_u x) = \mathcal{P}_r A \mathcal{P}_r x .$$

Let

$$y_n = \mathcal{P}_r x_n , \qquad n \geq 0 . \tag{4.2.10}$$

Then "by operating with" P_r on the left of the state equation in (4.2.1), we see that y_n satisfies:

$$y_n = (P_r A P_r) y_{n-1} + (P_r F) N_{n-1} ,$$

$$v_n = C y_n + G N_n . \tag{4.2.11}$$

Let

$$\hat{y}_n = E[y_n \mid v_n, \ldots, v_1] .$$

Then from (4.2.10) it follows that

$$\hat{y}_n = P_r \hat{x}_n .$$

But since $(CP_r \sim P_r A P_r)$ is observable, we see that

$$E[P_r x_n - P_r \hat{x}_n][P_r x_n - P_r \hat{x}_n]^* = E[(y_n - \hat{y}_n)(y_n - \hat{y}_n)]^*$$

satisfies (4.2.9). In other words,

$$P_r P_n P_r \leq M < \infty . \tag{4.2.12}$$

Next let

$$z_n = P_u A P_u z_{n-1} + P_u F N_{n-1} , \qquad n \geq 1 ; \tag{4.2.13}$$

$$z_0 = P_u x_0 .$$

Let us see how z_n differs from $P_u x_n$. By operating on the left with P_u on the state equation in (4.2.1), we have:

$$\begin{aligned} P_u x_n &= P_u A x_{n-1} + P_u F N_{n-1} \\ &= P_u A P_u x_{n-1} + P_u F N_{n-1} + P_u A P_r x_{n-1} \\ &= P_u A P_u x_{n-1} + P_u F N_{n-1} + P_u A y_{n-1} . \end{aligned} \tag{4.2.14}$$

Hence letting

$$h_n = \mathcal{P}_u x_n - z_n$$

and subtracting (4.2.14) from (4.2.13), we have

$$h_n = (\mathcal{P}_u A \mathcal{P}_u) h_{n-1} + \mathcal{P}_u A y_{n-1}$$

$$= \sum_0^{n-1} (\mathcal{P}_u A \mathcal{P}_u)^j \mathcal{P}_u A y_{n-1-j}$$

since

$$h_0 = \mathcal{P}_u x_0 - z_0 = 0 .$$

Now

$$x_n - \hat{x}_n = y_n - \hat{y}_n + z_n - \hat{z}_n + h_n - \hat{h}_n , \qquad (4.2.15)$$

where

$$\hat{z}_n = E[z_n \mid v_1, \ldots, v_n] ,$$

$$\hat{h}_n = E[h_n \mid v_1, \ldots, v_n] .$$

By virtue of the condition that unobservable states are stable, it follows that $(\mathcal{P}_u A \mathcal{P}_u)$ is stable. Hence

$$E[(z_n - \hat{z}_n)(z_n - \hat{z}_n)^*]$$

is bounded. Now for any two random vectors x, y we have

$$E[\|x+y\|^2] \leq E[\|x\|^2] + E[\|y\|^2] + 2E[\|x\| \, \|y\|] ,$$

and by the Schwarz inequality:

$$E[\|x\| \, \|y\|] \leq \sqrt{E[\|x\|^2]} \sqrt{E[\|y\|^2]} .$$

Hence

$$E[\|x+y\|^2] \leq \left(\sqrt{E[\|x\|^2]} + \sqrt{E[\|y\|^2]} \right)^2 .$$

By induction, it follows that for any finite number of random vectors $b_1, \ldots, b_n$, we have that

$$E\left[\left\| \sum_1^n b_i \right\|^2 \right] \leq \left(\sum_1^n \sqrt{E[\|b_i\|^2]} \right)^2 . \qquad (4.2.16)$$

Hence in particular

$$E[\|h_n - \hat{h}_n\|^2] \leq \left(\sum_0^{n-1} \sqrt{E[\|A_u^k P_u A(y_{n-1-k} - \hat{y}_{n-1-k})\|^2]} \right)^2$$

$$\leq \left(\sum_0^{n-1} \|A_u^k\| \, \|P_u A\| \sqrt{\text{Tr. } M} \right)^2 ,$$

where

$$A_u = P_u A P_u ,$$

$$M = \sup_k E[(y_k - \hat{y}_k)(y_k - \hat{y}_k)^*] .$$

Since A_u is stable, it follows therefore that

$$\sup_n E[\|h_n - \hat{h}_n\|^2] = \sup_n \text{Tr. } E[(h_n - \hat{h}_n)(h_n - \hat{h}_n)^*]$$

$$\leq \left(\sum_0^{\infty} \|A_u^k\| \, \|P_u A\| \sqrt{\text{Tr. } M} \right)^2$$

$$< \infty .$$

It follows in turn from (4.2.15) and the inequality (4.2.16) that

$$\sup_n E[\|x_n - \hat{x}_n\|^2] = \sup_n \text{Tr. } P_n$$

$$< \infty .$$

Or

$$\sup_n P_n < \infty ,$$

as was required to be proved.

Remark. It should be noted that h_n need not be zero, in general.

Lemma 4.2.3. Let P be any self-adjoint nonnegative definite solution of the steady state Riccati equation (4.2.7). If $(F^* \sim A^*)$ unobservable states are A^*-stable, then

$$(I - PC^*C)A$$

is stable, and moreover, P is the only self-adjoint nonnegative definite solution of (4.2.7).

Proof. Let P be a self-adjoint nonnegative definite solution of (4.2.7). Then

$$(I + H(P)C^*C)P = H(P) ,$$

or

$$P + H(P)C^*CP = H(P) ,$$

or

$$P = H(P)(I - C^*CP) ,$$

so that

$$P - PC^*CP = (I - PC^*C)P$$

$$= (I - PC^*C)H(P)(I - C^*CP) ,$$

or

$$P = (I - PC^*C)H(P)(I - C^*CP) + PC^*CP \qquad (4.2.17)$$

(which is no more than the steady-state version of (4.1.26)).

Let

$$\Psi = (I - PC^*C)A \quad ,$$

$$J = I - PC^*C \quad .$$

Then we can rewrite (4.2.17):

$$P = \Psi P \Psi^* + JFF^*J^* + PC^*CP \quad . \qquad (4.2.18)$$

Suppose Ψ is <u>not</u> stable. Then neither is Ψ^*. Let x be an unstable eigenvector of Ψ^*:

$$\Psi^* x = \gamma x \quad , \qquad |\gamma| \geq 1 \; , \quad ||x|| = 1 \quad .$$

Then substituting in (4.2.18), we have

$$[Px, x] = |\gamma|^2 [Px, x] + ||F^*J^*x||^2 + ||CPx||^2 \quad ,$$

or

$$(1 - |\gamma|^2)[Px, x] = ||F^*J^*x||^2 + ||CPx||^2 \quad . \qquad (4.2.19)$$

Since P is nonnegative definite,

$$[Px, x] \geq 0 \quad ,$$

and hence the right side of (4.2.19) must be zero if $|\gamma| \geq 1$. Hence

$$CPx = 0 \qquad \text{and} \qquad F^*J^*x = 0 \quad .$$

Since

$$J^*x = x + C^*CPx = x \quad ,$$

it follows that

$$F^*x = 0$$

and

$$\Psi^* x = A^* x$$
$$= \gamma x \ .$$

Hence for every nonnegative integer k:

$$F^*(A^*)^k x = \gamma^k F^* x = 0 \ ,$$

or x is $(F^* \sim A^*)$ Unobservable. This leads to a contradiction, since by assumption x must be A^*-stable. Hence Ψ must be stable.

Let us next prove uniqueness of the solution. Thus, let P_1, P_2 be two self-adjoint nonnegative definite solutions of the SSRE (4.2.7). Let

$$\Psi_1 = (I - P_1 C^* C)A \ ,$$
$$\Psi_2 = (I - P_2 C^* C)A \ .$$

Then both Ψ_1 and Ψ_2 must be stable. Let

$$Q = P_1 - P_2 \ .$$

As we have seen, if P is any solution of the SSRE, we have

$$(I + H(P)C^* C)^{-1} = I - C^* CP \ .$$

Hence the SSRE yields

$$H(P_1) = P_1(I - C^* CP_1)^{-1} \ ,$$
$$H(P_2) = (I - P_2 C^* C)^{-1} \ .$$

Subtracting, we have

$$AQA^* = P_1(I - C^*CP_1)^{-1} - (I - P_2C^*C)^{-1}P_2 \quad .$$

Hence

$$(I - P_2C^*C)AQA^*(I - C^*CP_1) = (I - P_2C^*C)P_1 - P_2(I - C^*CP_1)$$

$$= P_1 - P_2$$

$$= Q \quad ,$$

or

$$Q = \Psi_2 Q \Psi_1^* \quad ,$$

or for every positive integer k

$$Q = \Psi_2^k Q \Psi_1^{*k} \quad .$$

Hence

$$[Qx, x] = [Q \Psi_1^{*k} x, \Psi_2^{*k} x] \quad ,$$

and Ψ_1^*, Ψ_2^* being stable, letting k go to infinity, we obtain

$$[Qx, x] = 0 \quad .$$

Furthermore, Q being self-adjoint, it follows that

$$Q = 0$$

or

$$P_1 = P_2 \quad .$$

Remark. Note that for uniqueness we require that the solution (of the SSRE) be nonnegative definite. From (4.2.18) it follows that any self-adjoint solution is nonnegative definite if, in addition, Ψ is stable.

We can now complete the proof of Theorem 4.2.3 with the aid of our lemmas. First, the condition that (C-A) Unobservable states are A-stable yields that whatever P_0,

$$\sup_n P_n \leq M \leq \infty \quad ,$$

where

$$P_{n+1} = \Phi(P_n) \quad .$$

Take the special case where P_0 is zero. Then by Lemma 4.2.1

$$P_n = \Phi^n(0)$$

is monotone nondecreasing and, being bounded, converges to a finite limit. Denote the limit by P_s. Then P_s is a solution of the SSRE:

$$\Phi(P_s) = P_s \quad .$$

By the condition that $(F^* \sim A^*)$ Unobservable states are A^*-stable (Lemma 4.2.3) we obtain that any such solution must be unique. Moreover,

$$\Psi = (I - P_s C^* C)A$$

is stable. Let us consider the performance of the suboptimal filter, using P_s:

$$\left.\begin{aligned} \hat{x}^a_n &= A\hat{x}^a_{n-1} + P_s C^*(v_n - CA\hat{x}^a_{n-1}) \\ \hat{x}^a_0 &= 0 \end{aligned}\right\} \quad . \tag{4.2.20}$$

Let

$$e_n = x_n - \hat{x}^a_n \quad .$$

Let T_n denote the corresponding (error) covariance:

$$T_n = E[(x_n - \hat{x}_n^a)(x_n - \hat{x}_n^{a*})] .$$

Then T_n satisfies

$$T_{n+1} = \Psi T_n \Psi^* + (I - P_s C^* C)FF^*(I - P_s C^* C) + P_s C^* C P_s . \tag{4.2.20a}$$

But using the SSRE in the form (4.2.18), we see that

$$(T_{n+1} - P_s) = \Psi(T_n - P_s)\Psi^*$$

or

$$T_n - P_s = \Psi^n(\Lambda - P_s)\Psi^{*n} , \tag{4.2.21}$$

where

$$\Lambda = E[x_0 x_0^*] .$$

Since Ψ is stable, it follows that

$$\lim_n T_n = P_s ,$$

or the filter is asymptotically optimal, the rate of convergence to optimality being determined by (4.2.21). Clearly, the more stable Ψ the faster the convergence.

Next we note that for any initial P_0:

$$0 \leq P_0 .$$

Hence by Lemma 4.2. ,

$$\Phi^n(0) \leq \Phi^n(P_0) = P_n \leq T_n .$$

But the left side converges to P_s and the right side converges to P_s. Note, however, that the convergence need <u>not</u> be monotone.

Remark 1. Note that (4.2.20) defines a Gaussian process which is asymptotically stationary and ergodic. In particular, we have that

$$\lim_{N\to\infty} \frac{1}{N} \sum_1^N e_n e_n^* = P_s .$$

Remark 2. Note that if A is stable, all the conditions of Theorem 4.2.3 are automatically satisfied.

One-Dimensional Example

The conditions of Theorem 4.2.3 are necessary for the conclusions therein to hold. Let us specialize to the simplest possible case for this purpose. Thus, let the state and observation be both one-dimensional, and let

$$\left.\begin{aligned} v_n &= s_n + gN_n^o \\ s_n &= cx_n \\ x_n &= \rho x_{n-1} + fN_n^s \end{aligned}\right\} , \qquad (4.2.1E)$$

where $\{N_n^o\}$ and $\{N_n^s\}$ are mutually independent white-noise sequences, each with unit variance; and ρ is allowed to be any real number, and we use p for P. Without loss of generality, we may, and do, take g to be unity. Then

$$\phi(p) = \frac{|\rho|^2 p + f^2}{1 + c^2(|\rho|^2 p + f^2)} ,$$

$$\phi'(p) = \frac{|\rho|^2}{(1 + c^2(|\rho|^2 p + f^2))^2} ,$$

$$\phi''(p) = \frac{-2|\rho|^4 c^4}{(1 + c^2(|\rho|^2 p + f^2))^3} .$$

Thus $\phi(p)$ is a convex, increasing function of p for $0 \le p < \infty$, degenerating to a straight line for $c = 0$. See Figure 4.1.

Note that

$$p_{n+1} = \phi(p_n)$$

is always monotone:

$$\text{increasing if } p_1 \ge p_0 ,$$

$$\text{decreasing if } p_1 \le p_0 .$$

The SSRE

$$\phi(p) = p$$

is a quadratic equation in p and has two "real" solutions: one positive and one negative, the positive solution being

$$p_\infty = \frac{-(1-|\rho|^2+c^2f^2) + \sqrt{(1-|\rho|^2+c^2f^2)^2 + 4|\rho|^2c^2f^2}}{2|\rho|^2c^2} , \qquad (4.2.2E)$$

and this formula, of course, requires appropriate interpretation if $c = 0$ or $f = 0$.

<u>Case 1</u>. $c = 0$; $f \neq 0$ (Controllable but <u>not</u> Observable).

In this case the graph of $\phi(p)$ degenerates to a straight line and the SSRE degenerates to the linear equation

$$0 = f^2 - (1 - |\rho|^2)p$$

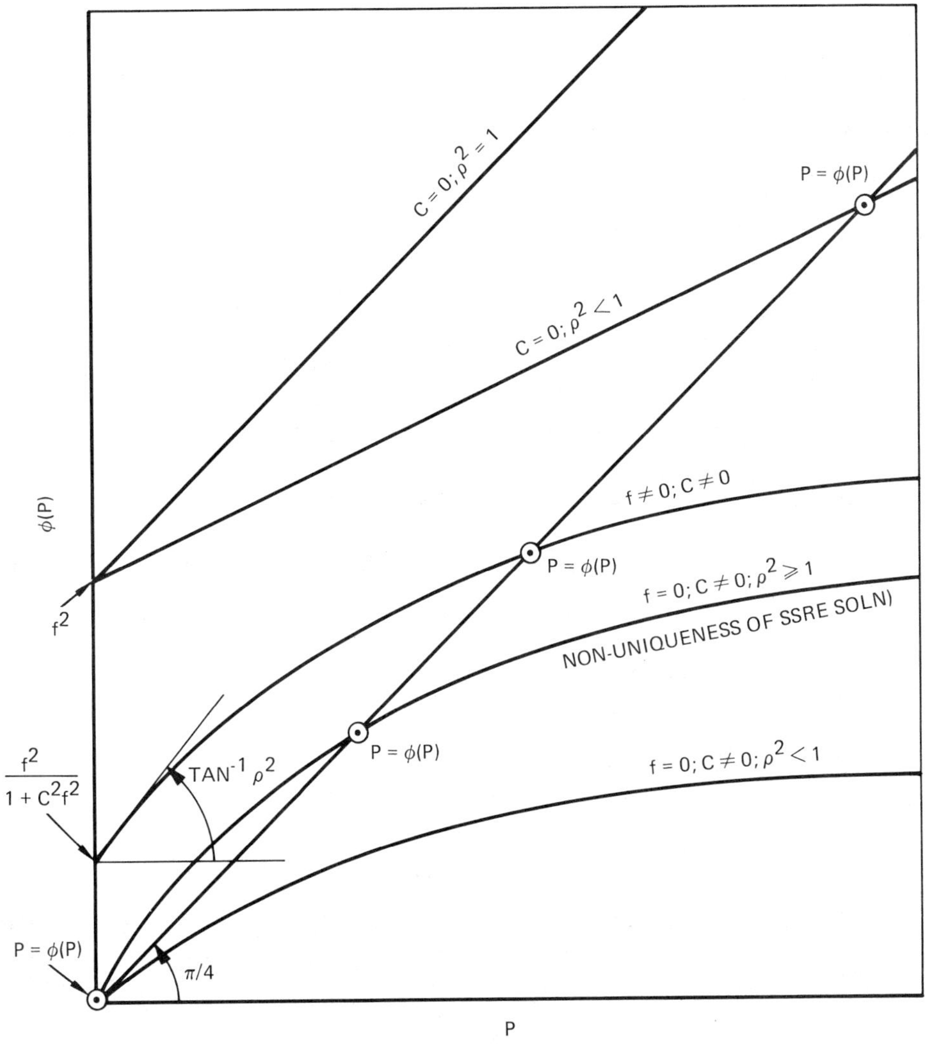

FIGURE 4.1. Behaviour of $\phi(P)$: One Dimension.

and thus has a unique nonnegative solution

$$p_\infty = \frac{f^2}{1 - |\rho|^2}$$

if and only if $|\rho| < 1$ (system stable). These facts are also deducible from the sketch in Figure 4.1. If $c = 0$ and $|\rho| > 1$ (Unobservable states are <u>not</u> stable), we see that p_n goes to infinity.

<u>Case 2.</u> $f = 0$; $c \neq 0$ (Observable but <u>not</u> Controllable).

In this case the SSRE has always the nonnegative solution

$$p_\infty = 0 \quad ,$$

which is unique if $|\rho| < 1$; however, if $|\rho| > 1$ (Uncontrollable states are <u>not</u> stable), it has the additional solution

$$p = \frac{1 - |\rho|^{-2}}{c^2} \quad , \qquad |\rho| > 1 \ ;$$

and only this solution will make

$$(1 - p_\infty c^2)\rho$$

stable.

<u>Case 3</u>. $f = 0$; $c = 0$ (Neither Observable nor Controllable).

The SSRE becomes

$$0 = (1 - |\rho|^2)p \quad ,$$

which has only one solution:

$$p_\infty = 0$$

so that

$$(1 - p_\infty c^2)\rho = \rho$$

(or in this case the filter degenerates to a "direct connection," $\hat{s}_n = s_n$).

Miscellaneous Remarks

It is evident that these conclusions are consistent with our general asymptotic theory. We can also study what happens when one or both of the assumptions of Theorem 4.2.3 are violated in the multidimensional case. In particular, the case where (A-F) is not Controllable is of interest to us in practical filter implementation, especially in the case where A is stable. Our first result in this connection is:

Theorem 4.2.4. Suppose A is stable. Then

$$P_\infty x = 0 \tag{4.2.22}$$

for all $(F^* \sim A^*)$ Unobservable states, that is, for all x such that

$$F^* A^{*k} x = 0 \quad , \qquad k \geq 0 \quad , \tag{4.2.23}$$

and, conversely, (4.2.22) implies (4.2.23).

Proof. Since A is stable, Theorem 4.2.3 applies so that P_∞ is well-defined as the unique solution of the SSRE. Let

$$R_n = E[x_n x_n^*] \quad .$$

Then

$$R_{n+1} = AR_n A^* + FF^* \quad ,$$

and since A is stable, R_n converges to

$$R_\infty = \sum_0^\infty A^k FF^* A^{*k} \quad .$$

Now the null space of R_∞ is readily seen to be the class of $(F^* \sim A^*)$ Unobservable states. On the other hand

$$P_n \leq R_n \quad ,$$

and P_n converges to P_∞ so that

$$[P_\infty x, x] \leq [R_\infty x, x] \quad .$$

Hence for x satisfying (4.2.23)

$$P_\infty x = 0 \quad .$$

Conversely, suppose (4.2.22) holds. Then the SSRE implies that

$$H(P_\infty)x = AP_\infty A^* x + FF^* x = 0 \quad ,$$

or

$$F^* x = 0 \qquad \text{and} \qquad P_\infty A^* x = 0 \quad .$$

The second equality says that $A^* x$ satisfies (4.2.22). Hence

$$F^* A^* x = 0 \quad ,$$

and by induction (4.2.23) follows.

The singularity of P_∞ is undesirable from the practical point of view. Suppose, for example, we use the asymptotically optimal filter defined by (4.2.20). Then we see that for x satisfying (4.2.22)

$$[\hat{x}_n^a, x] = [\hat{x}_{n-1}^a, A^* x]$$

since

$$[P_\infty C^* [v_n - CA\hat{x}^a_{n-1}], x] = 0 .$$

Hence, for <u>all</u> n

$$[\hat{x}^a_n, x] = 0 \qquad \text{since} \quad \hat{x}^a_0 = 0 ,$$

or we ignore any information contained in the observation concerning $(F^* \sim A^*)$ Unobservable states. In this case we may construct a filter which, while suboptimal, does not ignore this information. For this purpose, we proceed as follows. Define the new SSRE:

$$P = (I - PC^*C)(APA^* + \sigma^2 I + FF^*) \tag{4.2.24}$$

corresponding to replacing FF^* by

$$\sigma^2 I + FF^* ,$$

or replacing F by $\sqrt{\sigma^2 I + FF^*}$. The latter is nonsingular, and hence

$$\left(A \sim \sqrt{\sigma^2 I + FF^*}\right)$$

is Controllable; and A being stable, (4.2.24) has a unique nonnegative definite solution. Denote it by P_σ. Then P_σ is nonsingular, and P_σ converges to P_∞ as $\sigma \to 0$. We may therefore use P_σ in place of P_∞, and while the filter would be suboptimal, the difference in performance would be small for small enough σ. In fact, let

$$\left.\begin{aligned} \hat{x}^\sigma_n &= A\hat{x}^\sigma_{n-1} + P_\sigma C^* [v_n - CA\hat{x}^\sigma_{n-1}] \\ \hat{x}^\sigma_0 &= 0 \end{aligned}\right\} . \tag{4.2.25}$$

Let

$$z_n = x_n - \hat{x}_n^\sigma ;$$

$$T_n = E[z_n z_n^*] .$$

Then, as before (cf. (4.2.20a)), we have

$$T_{n+1} = \Psi_\sigma T_n \Psi_\sigma^* + P_\sigma C^* C P_\sigma + (I - P_\sigma C^* C) FF^* (I - C^* C P_\sigma) ,$$

where

$$\Psi_\sigma = (I - P_\sigma C^* C) A .$$

Since Ψ_σ is stable, T_n converges to T_∞, say:

$$T_\infty = \Psi_\sigma T_\infty \Psi_\sigma^* + Q_\sigma ,$$

where

$$Q_\sigma = P_\sigma C^* C P_\sigma + (I - P_\sigma C^* C) FF^* (I - C^* C P_\sigma)$$

$$= P_\sigma - \Psi_\sigma P_\sigma \Psi_\sigma^* - \sigma^2 (I - P_\sigma C^* C)(I - C^* C P_\sigma) ,$$

since we can rewrite (4.2.24) as (cf. (4.2.17)):

$$P_\sigma = \Psi_\sigma P_\sigma \Psi_\sigma^* + P_\sigma C^* C P_\sigma$$
$$+ (I - P_\sigma C^* C)(\sigma^2 I + FF^*)(I - C^* C P_\sigma) . \quad (4.2.24a)$$

Hence

$$T_\infty - P_\sigma = \Psi_\sigma (T_\infty - P_\sigma) \Psi^* - \sigma^2 (I - P_\sigma C^* C)(I - P_\sigma C^* C)^*$$

or

$$T_\infty - P_\sigma = -\sigma^2 \sum_0^\infty \Psi_\sigma^k (I - P_\sigma C^* C)(I - P_\sigma C^* C)^* \Psi_\sigma^{*k} . \quad (4.2.26)$$

Let P_0 denote P_σ at $\sigma = 0$:

$$P_0 = (I - P_0 C^* C)(A P_0 \Psi^* + FF^*) .$$

Also let

$$\Psi_0 = (I - P_0 C^* C)A \quad .$$

Let us examine the behavior of P_σ as σ goes to zero. We have

$$AP_\sigma A^* + \sigma^2 I + FF^* = P_\sigma (I - C^* CP_\sigma)^{-1}$$

$$AP_0 A^* + FF^* = (I - P_0 C^* C)^{-1} P_0 \quad .$$

Hence (as in proving the uniqueness of solution to the SSRE) we have

$$\Psi_0 (P_\sigma - P_0) \Psi_\sigma^* = -\sigma^2 (I - P_0 C^* C)(I - C^* CP_\sigma) + P_\sigma - P_0 \quad .$$

Since Ψ_0 and Ψ_σ are stable, it follows that

$$P_\sigma - P_0 = \sigma^2 \sum_0^\infty \Psi_0^k (I - P_0 C^* C)(I - C^* CP_\sigma) \Psi_\sigma^{*k} \quad . \tag{4.2.27}$$

Hence, adding (4.2.26) and (4.2.27), we have

$$T_\infty - P_0 = \sigma^2 \sum_0^\infty \Big(\Psi_0^k (I - P_0 C^* C)(I - C^* CP_\sigma) \Psi_\sigma^{*k} - \Psi_\sigma^k (I - P_\sigma C^* C)(I - C^* CP_\sigma) \Psi_\sigma^{*k} \Big) . \tag{4.2.28}$$

This shows, in particular, that

$$\frac{T_\infty - P_0}{\sigma^2} \to 0 \qquad \text{as} \quad \sigma \to 0 \quad ,$$

so that the suboptimal filter performance may still be acceptable for small σ.

Finally let us enumerate equivalent alternate forms of

the SSRE.

$$P = (I - PC^*C)APA^*(I - PC^*C) + (I - PC^*C)FF^*(I - PC^*C)^* + PC^*CP \quad , \qquad (4.2.29)$$

$$P = (I + H(P)C^*C)^{-1}H(P) \quad , \qquad (4.2.30)$$

$$P^{-1} = (H(P)^{-1} + C^*C) \quad \text{(if } P \text{ is nonsingular)} \; , \qquad (4.2.30a)$$

$$P = (I - PC^*C)H(P) \quad , \qquad (4.2.31)$$

$$P = APA^* + FF^* - PC^*(I + CH(P)C^*)CP$$
$$\text{(steady-state version of (4.1.24))}. \qquad (4.2.32)$$

(Note that

$$FF^* - PC^*(I + CH(P)C^*)CP$$

is <u>not</u> necessarily (positive) definite.) From (4.2.30) we can deduce that P must satisfy:

$$CPC^* = (I + CH(P)C^*)^{-1} CH(P)C^* \qquad (4.2.33)$$

as follows:

$$(I + CH(P)C^*)CPC^* = C(I + H(P)C^*C)PC^*$$

$$= CH(P)C^*$$

by (4.2.30). However, from (4.2.33) we can only obtain that

$$0 = (I + CH(P)C^*)CPC^* - CH(P)C^*$$

$$= C((I + H(P)C^*C)P - H(P))C^*$$

$$= 0 \quad ,$$

or that

$$(I + H(P)C^*C)P - H(P) = Z ,$$

where

$$CZC^* = 0 .$$

In other words, (4.2.33) is <u>not</u> necessarily <u>equivalent</u> to (4.2.30). Similarly we can also deduce from (4.2.31) that

$$CPC^* = (I - CPC^*)CH(P)C^* \tag{4.2.34}$$

as follows

$$(I - CPC^*)CH(P)C^* = C(I - PC^*C)H(P)C^*$$

$$= CPC^*$$

by (4.2.31). In particular, we also have (cf. (4.1.28) also):

$$(I + CH(P)C^*)^{-1} = I - CPC^* .$$

★ PROBLEMS ★

<u>Problem 4.2.1</u>

Let R be a covariance and A be stable. Then

$$ARA^* \leq R$$

if and only if R has the form

$$R = \sum_0^\infty A^n Q A^{*n} , \qquad Q \geq 0 .$$

Construct a stable matrix A such that

$$I - AA^*$$

is not a covariance.

Hint: Take

$$A = \begin{bmatrix} 0 & b \\ a & 0 \end{bmatrix} ,$$

where $b > 1$, $a < 1$ and $ab < 1$.

Problem 4.2.2

Show that if $(A \sim F)$ is Controllable then $H(P)$ and P are nonsingular, P being any self-adjoint nonnegative solution of the SSRE. Show that $(A \sim F)$ is not Controllable if P is singular. Show that if A is stable and P is singular, then $(A \sim F)$ cannot be Controllable. Let $\mathcal{N}(C)$ denote the null space of C. Show that $(I - PC^*C)$ has an eigenvalue equal to unity if and only if $\mathcal{N}(C)$ is nonzero.

Problem 4.2.3

Let P be the unique solution of the SSRE under the conditions of Theorem 4.2.3. Show that

$$P \geq (I - PC^*C)P(I - C^*CP) + PC^*CP .$$

Hint: Use Problem 4.1.8.

Problem 4.2.4

Show that for P, Q self-adjoint and nonnegative definite:

$$f(\lambda) = \Phi(P + \lambda Q)$$

is a convex function of λ, $-\infty < \lambda < \infty$. Recall that $\Phi(\cdot)$ is defined in (4.2.6).

Hint:

$$\frac{d^2 f(\lambda)}{d\lambda^2} = -h(\lambda)AQA(h(\lambda)^* C^* C + C^* Ch(\lambda))AQA^* h(\lambda)^*,$$

where

$$h(\lambda) = (I - \Phi(P+\lambda Q)C^* C)$$

$$= (I + H(P+\lambda Q)C^* C)^{-1}$$

and $C^* Ch(\lambda)$ is self-adjoint and nonnegative definite.

4.3. STEADY STATE THEORY: FREQUENCY DOMAIN ANALYSIS

Prior to the advent of digital computers, filtering theory was primarily studied in the frequency domain rather than the time domain. It is instructive to study Kalman filter design -- and particularly the steady state properties of the filter -- from this point of view.

One-Dimensional Example

Let us begin by considering the one-dimensional case, using the notation of (4.2.1E) of Section 4.2.2:

$$s_n = cx_n , \tag{4.3.1}$$

$$\left. \begin{aligned} x_{n+1} &= \rho x_n + fN_n^s \\ v_n &= s_n + N_n^o \end{aligned} \right\} . \tag{4.3.2}$$

Since, in the steady state, we can solve (4.3.2) to yield

$$s_n = cf \sum_0^\infty \rho^k N^s_{n-k-1} ,$$

the "signal" "transfer function" (cf. Chapter 2), denoted $\psi_s(\lambda)$, is

$$\psi_s(\lambda) = \frac{cf}{e^{2\pi i\lambda} - \rho} , \qquad -\tfrac{1}{2} < \lambda < \tfrac{1}{2} ; \tag{4.3.3}$$

and the spectral density of the signal process is

$$p_s(\lambda) = \frac{c^2 f^2}{|e^{2\rho i\lambda} - \rho|^2} = \frac{c^2 f^2}{1 + \rho^2 - 2\rho \cos 2\pi\lambda} . \tag{4.3.4}$$

The steady state Kalman filter is

$$\hat{x}_n = (1 - pc^2)\rho\hat{x}_{n-1} + pcv_n ; \tag{4.3.5}$$

$$\hat{s}_n = cx_n .$$

Hence the corresponding "transfer function" is

$$\psi_f(\lambda) = \frac{pc^2}{e^{2\pi i\lambda} - \rho_F} , \tag{4.3.6}$$

where

$$\rho_F = (1 - pc^2)\rho .$$

Since

$$\rho_F \leq \rho ,$$

we have immediately that for the "normalized" transfer functions:

$$\frac{(1 - \rho_F)}{|e^{2\pi i\lambda} - \rho_F|} \geq \frac{(1 - \rho)}{|e^{2\pi i\lambda} - \rho|} , \qquad -\tfrac{1}{2} \leq \lambda \leq \tfrac{1}{2}. \tag{4.3.7}$$

This means that the "gain frequency" curve of the Kalman filter defined by

$$\log |e^{2\pi i\lambda} - \rho_F| \quad , \qquad -\tfrac{1}{2} \leq \lambda \leq \tfrac{1}{2} \tag{4.3.8}$$

is always "flatter" than that of the signal filter

$$\log |e^{2\pi i\lambda} - \rho| \quad , \qquad -\tfrac{1}{2} \leq \lambda \leq \tfrac{1}{2} \, . \tag{4.3.9}$$

This (following tradition) can be put in another way. The half-power ("3db point") "break" frequency λ of any transfer function is defined to be that value of λ for which the magnitude is half the maximum. Hence for the Kalman filter it is defined by

$$\frac{|1 - \rho_F|^2}{|e^{2\pi i\lambda} - \rho_F|^2} = \frac{1}{2} \; ,$$

or λ is given by

$$\frac{1}{\pi} \sin^{-1}\left(\frac{1 - \rho_F}{2\sqrt{\rho_F}}\right), \qquad \left(\approx \frac{|\log \rho_F|}{2\pi} \text{ for } \rho_F \text{ close to } 1\right),$$

which is larger than that of the signal transfer function break frequency which is

$$\frac{1}{\pi} \sin^{-1}\left(\frac{1 - \rho}{2\sqrt{\rho}}\right), \qquad \left(\approx \frac{|\log \rho|}{2\pi} \text{ for } \rho \text{ close to } 1\right).$$

Of special interest to us is the dependence of the Kalman filter transfer function on signal-to-noise ratio (S/N):

$$\frac{\text{Signal Power}}{\text{(Observation) Noise Power}} \; , \tag{4.3.10}$$

which in the one-dimensional case is

$$= \frac{c^2 f^2}{1 - \rho^2} \quad .$$

Large cf (equivalently, large c, $f \neq 0$) corresponds to large signal (or low noise), and small cf ($\neq 0$) corresponds to small signal (or large noise). Let us therefore consider the limiting cases as $c \to \infty$ and $c \to 0$ (the latter is referred to as the "threshold" case). We can readily deduce from (4.2.2E) that:

Suppose $c \to \infty$; $f \neq 0$; $\rho > 1$.

Then $pc \to 0$; $pc^2 \to 1$; $c\rho_F \to 0$; $s_n \approx v_n$; $E(s_n - \hat{s}_n)^2 \approx 1$;

but the "normalized" error

$$\frac{E((s_n - \hat{s}_n)^2)}{E((s_n)^2)} \approx 0 \quad .$$

Suppose $c \to 0$. Then $pc^2 \to 0$; $\rho_F \to \rho$; $c\rho_F \to 0$; $\hat{x}_n \approx \hat{x}_{n-1}$ (or $\hat{x}_n = 0$ asymptotically); $\hat{s}_n \approx 0$.

In terms of the Kalman filter transfer function we thus see that

(a) For low noise or large signal, the filter allows all frequencies to pass without attenuation, or simply "stops filtering": $\rho_F \approx 0$ ($\hat{s}_n \approx v_n$).

(b) For the threshold case of large noise, the Kalman filter transfer function "filters" out the noise as much as it can, and its gain frequency curve approaches that of the signal transfer function. See Figure 4.2.

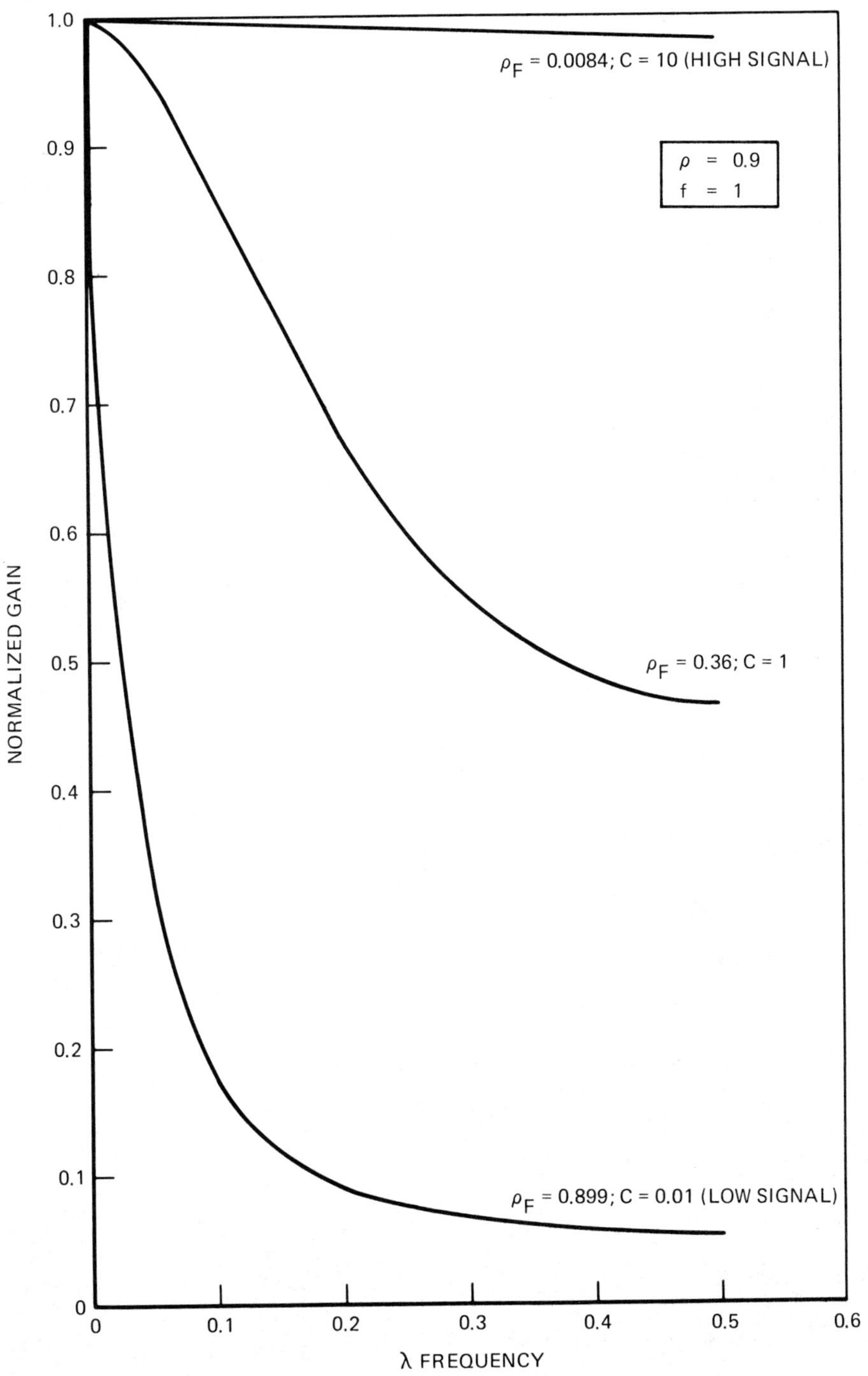

FIGURE 4.2. Normalized K.F. Gain: One Dimension (Low Pass).

Two-Dimensional (Two-State) Example

Next let us consider the "band-pass" case (Linear Oscillator: cf. Chapter 2), as opposed to the "low-pass" case as in our first example. To enable explicit calculation, we take $\omega_0\Delta$ therein to be $\frac{\pi}{2}$, and $b = 0$ (no damping); so that we have the model (omitting constants):

$$x_{n+1} = Ax_n + FN_n \quad ;$$

$$v_n = Cx_n + GN_n \quad ,$$

where

$$A = \begin{bmatrix} 0 & +1 \\ -1 & 0 \end{bmatrix} \quad ;$$

$$F = \begin{vmatrix} 0 & 0 \\ 1 & 0 \end{vmatrix} \quad ;$$

$$G = | \, 0 \quad 1 \, | \quad ;$$

$$N_n = \begin{vmatrix} N_n^s \\ N_n^0 \end{vmatrix} \quad ,$$

white noise with unit variance. To study the effect of signal-noise ratio we shall take

$$C = | \, \gamma, \; 0 \, | \quad ,$$

where $0 < \gamma < \infty$. The matrix A has its eigenvalue on the unit circle and allowing for singularities, the signal normalized gain frequency function is

$$= \frac{2}{|1 + e^{4\pi i\lambda}|} \quad . \qquad (4.3.11)$$

The singularities are at

$$\lambda = \pm \frac{1}{4} \quad .$$

We may consider the signal as arising from white noise through an oscillatory or band-pass system with low damping. We know that the Kalman filter must be stable, and hence it is of interest to study its transfer function. The steady state Riccati equation is readily verified to have the solution

$$P_\gamma = \begin{bmatrix} \frac{1}{2}\left(\sqrt{1 + \frac{4}{\gamma^2}} - 1\right) & 0 \\ 0 & \frac{1}{2}\left(\sqrt{1 + \frac{4}{\gamma^2}} + 1\right) \end{bmatrix} , \qquad (4.3.12)$$

where the subscript γ is to indicate dependence on γ. Hence

$$\begin{aligned} \psi_\gamma &= (I - P_\gamma C^* C)A \\ &= \begin{bmatrix} 0 & 1 - \frac{\gamma^2}{2}\left(\sqrt{1 + \frac{4}{\gamma^2}} - 1\right) \\ -1 & 0 \end{bmatrix} , \end{aligned}$$

and the Kalman filter transfer function is:

$$C[e^{2\pi i\lambda} I - \psi_\gamma]^{-1} P_\gamma C^* \quad .$$

Normalizing this, by dividing it by the value at $\lambda = 0$, we obtain for the normalized transfer function

$$g(\lambda) = \frac{e^{2\pi i\lambda}(1+a)}{a + e^{4\pi i\lambda}} ,$$

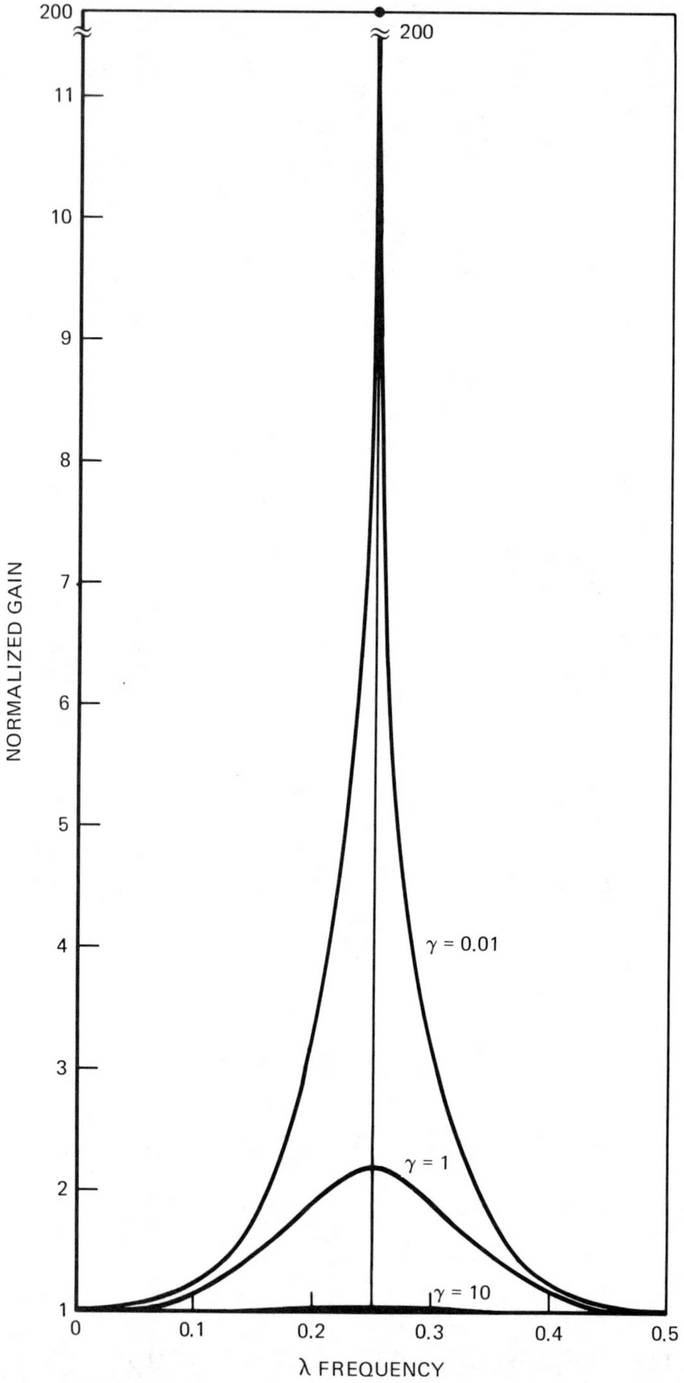

FIGURE 4.3. K. F. Transfer Gain: Band-Pass Case.

where

$$a = 1 - \frac{\gamma^2}{2}\left(\sqrt{1+\frac{4}{\gamma^2}} - 1\right) .$$

Note that

$$0 < a < 1$$

and $|g(\lambda)|$ has its maximum at

$$\lambda = \pm\frac{1}{4} ,$$

at the same points where the signal transfer function has singularities, but now

$$\max |g(\lambda)| = \frac{1+a}{1-a} < \infty .$$

See Figure 4.3, where the graph of $|g(\lambda)|$ is sketched for $a < 1$ and $a = 1$. Note that unlike the "low-pass" case, the gain frequency curve of the Kalman filter lies always below that of the signal filter. Note also that

$$a \to 0 \quad \text{as} \quad \gamma \to \infty ;$$

$$a \to 1 \quad \text{as} \quad \gamma \to 0 ,$$

so that we may make the same conclusions as in the previous one-dimensional case for the Kalman filter behavior at high and low signal-noise ratios. We omit the details.

General Case

Let us now go on to the general case. We shall assume that A is stable (the "marginally" stable case when the eigenvalues are allowed to be on the unit circle can be handled allowing for "singularities" of the transfer function). Also we shall assume ($A \sim F$) controllability and ($C \sim A$) observability to assure stability of the steady state Kalman filter. Let

$$\psi = (I - PC^*C)A \quad .$$

Then the Kalman filter transfer function (matrix) is:

$$g(\lambda) = C(I - \psi e^{2\pi i\lambda})^{-1} PC^* \quad , \tag{4.3.13}$$

as opposed to the signal transfer function

$$f(\lambda) = C(I - Ae^{2\pi i\lambda})^{-1} F \quad . \tag{4.3.14}$$

Let the eigenvalues of ψ be $\{\rho_i\}$. Then

$$\begin{aligned} e^{2\pi i\lambda}(I - \psi e^{2\pi i\lambda})^{-1} &= (e^{-2\pi i\lambda} - \psi)^{-1} \\ &= \left\{ \frac{b_{ij}(e^{-2\pi i\lambda})}{\prod_k (e^{-2\pi i\lambda} - \rho_k)} \right\} \quad , \end{aligned} \tag{4.3.15}$$

where $b_{ij}(\cdot)$ is a polynomial of degree less than of the denominator and hence we can use partial fraction expansion. We know that $|\rho_i| < 1$ and, if complex, must come in conjugate pairs. We may thus obtain sums of terms each corresponding to one of the two examples we have described above. We omit the details.

Let us now turn to the limiting cases of low noise and high noise. It is convenient for this purpose to take the noise variance matrix

$$GG^* = \gamma I \quad .$$

Then the Kalman filter state equations are

$$\hat{s}_n = C\hat{x}_n \quad ;$$

$$\hat{x}_{n+1} = \left(I - P_\gamma \frac{C^*C}{\gamma}\right)A\hat{x}_n + P_\gamma \frac{C^*}{\gamma} v_n \quad ;$$

and

$$P_\gamma = \left(I + \frac{H(P_\gamma)C^*C}{\gamma}\right)^{-1} H(P_\gamma) \quad .$$

First of all consider the suboptimal estimate

$$\hat{s}_n^a = v_n \quad .$$

The corresponding error matrix is clearly γI and of course,

$$CP_\gamma C^* \leq \gamma I \quad . \tag{4.3.16}$$

In particular, the error covariance for the suboptimal filter goes to zero as γ goes to zero, or it is optimal in the limit. Hence we should expect that

$$\frac{CP_\gamma C^*}{\gamma} v \rightarrow v$$

on $\mathcal{R}(C)$, the range space of C, and that

$$C\left(I - \frac{P_\gamma C^*C}{\gamma}\right) \rightarrow 0$$

as $\gamma \rightarrow 0$. Since the second limit follows from the first, we need only prove the first. Now we know that

$$\frac{CP_\gamma C^*}{\gamma} = \left(I + C\frac{H(P_\gamma)}{\gamma}C^*\right)^{-1} C\frac{H(P_\gamma)}{\gamma}C^* \quad . \tag{4.3.17}$$

To proceed, let

$$Q_\gamma = \frac{P_\gamma}{\gamma} \quad ;$$

$$K_\gamma = AQ_\gamma A^* + \frac{FF^*}{\gamma} \quad ;$$

$$T_\gamma = CQ_\gamma C^* \quad ;$$

$$H_\gamma = CK_\gamma C^* \quad .$$

Then we have from (4.3.17) that

$$I - T_\gamma = (I + H_\gamma)^{-1}$$

and, of course,

$$T_\gamma \leq I \quad .$$

Let us examine more closely how T_γ depends on γ. For this purpose, let us differentiate

$$Q_\gamma = (I + K_\gamma C^* C)^{-1} K_\gamma$$

with respect to γ. Or, equivalently,

$$Q_\gamma^{-1} = (K_\gamma^{-1} + C^* C)$$

yielding

$$-Q_\gamma^{-1} \frac{dQ_\gamma}{d\gamma} Q_\gamma^{-1} = -K_\gamma^{-1}\left(A \frac{dQ_\gamma}{d\gamma} A^* - \frac{FF^*}{\gamma^2}\right)K_\gamma^{-1} \quad ,$$

or

$$\frac{dQ_\gamma}{d\gamma} = J_\gamma A \frac{dQ_\gamma}{d\gamma} (J_\gamma A)^* - J_\gamma \frac{FF^*}{\gamma^2} J_\gamma^* \quad , \qquad (4.3.18)$$

where

$$J_\gamma = (I - Q_\gamma C^* C) \quad ,$$

since

$$Q_\gamma K_\gamma^{-1} = I - Q_\gamma C^* C \quad .$$

Now from

$$P_\gamma = J_\gamma A P_\gamma (J_\gamma A)^* + J_\gamma F F^* J_\gamma^* + P_\gamma C^* C Q_\gamma$$

we have that

$$Q_\gamma = (J_\gamma A) Q_\gamma (J_\gamma A)^* + J_\gamma \frac{FF^*}{\gamma} J_\gamma^* + Q_\gamma C^* C Q_\gamma \quad . \tag{4.3.19}$$

We can prove from this (just as in Lemma 4.2.3) that $J_\gamma A$ is stable, since

$$\left(A \sim \frac{F}{\sqrt{\gamma}} \right)$$

is controllable. Hence we obtain from (4.3.18) that

$$\frac{dQ_\gamma}{d\gamma} = - \sum_0^\infty (J_\gamma A)^k \frac{J_\gamma F F^* J_\gamma^*}{\gamma^2} (J_\gamma A)^{*k} \quad , \tag{4.3.20}$$

from which it follows that Q_γ increases as γ decreases. In particular, then T_γ is monotone nondecreasing as γ decreases, and since it is bounded, must converge to, say T_0. Let us now prove that

$$T_0 = I \quad \text{on} \quad \mathcal{R}(C) \quad .$$

For this purpose, we solve (4.3.19), obtaining

$$Q_\gamma = \sum_0^\infty (J_\gamma A)^k \left(\frac{J_\gamma F F^* J_\gamma^*}{\gamma} + Q_\gamma C^* C Q_\gamma \right) (J_\gamma A)^{*k} \quad .$$

Hence

$$T_\gamma = \sum_0^\infty C (J_\gamma A)^k \frac{J_\gamma F F^* J_\gamma^*}{\gamma} (J_\gamma A)^{*k} C^*$$

$$+ \sum_0^\infty C (J_\gamma A)^k Q_\gamma C^* C Q_\gamma (J_\gamma A)^{*k} C^* \quad .$$

In the second sum the term for $k = 0$ is clearly T_γ^2, and hence

$$T_\gamma - T_\gamma^2 = \sum_0^\infty C(J_\gamma A)^k \frac{J_\gamma FF^* J_\gamma^*}{\gamma} (J_\gamma A)^{*k} C^*$$

$$+ \sum_1^\infty C(J_\gamma A)^k Q_\gamma C^* C Q_\gamma (J_\gamma A)^{*k} C^* . \quad (4.3.21)$$

Next let $\gamma \to 0$. The left side converges and hence so must the right side; in fact, each term in each summation must converge, being nonnegative. Hence for each v,

$$\frac{\|F^* J_\gamma^* (J_\gamma A)^{*k} C^* v\|^2}{\gamma} \quad \text{converges} \quad ;$$

$$\|C Q_\gamma (J_\gamma A)^{*k} C^* v\|^2 \quad \text{converges} \quad .$$

The idea is now to deduce from this that

$$F^* A^{*k} C^* (I - T_0) v = 0 \quad (4.3.22)$$

for every k, and by (A-F) Controllability it will then follow that

$$C^* (I - T_0) v = 0 \quad ;$$

and since v and $T_0 v$ are in the range of C, it will follow further that

$$T_0 v = v \quad \text{on} \quad \mathcal{R}(C) \quad ,$$

as required.

Let us next deduce (4.3.22). Taking $k = 0$ in the first summation in (4.3.21), we have that

$$\|F^* J_\gamma^* C^* v\| = \|F^* C^* (I - T_\gamma) v\| \quad ,$$

and hence

$$\frac{||F^*C^*(I - T_\gamma)v||}{\gamma} \quad \text{is bounded} \quad ;$$

and hence

$$||F^*C^*(I - T_\gamma)v|| \quad \text{must converge to} \quad 0$$

and hence

$$F^*C^*(I - T_0)v \;=\; 0 \;. \tag{4.3.23}$$

Taking $k = 1$ in the second sum in (4.3.21), we have that

$$||CQ_\gamma A^*(I - C^*CQ_\gamma)C^*v|| \;=\; ||CQ_\gamma A^*C^*(I - T_\gamma)v|| \tag{4.3.24}$$

converges; while taking $k = 1$ in the first sum in (4.3.21) we have

$$F^*(I - C^*CQ_\gamma)A^*(I - C^*CQ_\gamma)C^*v$$

$$= \; F^*(I - C^*CQ_\gamma)A^*C^*(I - T_\gamma)v$$

$$= \; F^*A^*C^*(I - T_\gamma)v \; - \; F^*C^*CQ_\gamma A^*C^*(I - T_\gamma)v \quad ;$$

and by the convergence of (4.3.24) it follows that so does

$$F^*C^*CQ_\gamma A^*C^*(I - T_\gamma)v \quad ,$$

and hence so does

$$\frac{||F^*A^*C^*(I - T_\gamma)v||}{\sqrt{\gamma}} \quad .$$

Hence

$$F^*A^*C^*(I - T_0)v \;=\; 0 \;.$$

We may continue in this manner to obtain (4.3.22), successively for each k, $k \leq n-1$, where A is $n \times n$.

The case of low signal-noise ratio or high noise is much simpler. Thus we take the system as

$$x_n = Ax_{n-1} + FN_{n-1} \quad ;$$

$$v_n = Cx_n + N_n \quad ,$$

where

$$GG^* = \gamma I \quad ,$$

and allow γ to go to infinity. Now the suboptimal signal estimate

$$\hat{s}_n^a = 0$$

has the error covariance CRC^*, and hence

$$CP_\gamma C^* \leq CRC^* \quad ,$$

or

$$\frac{CP_\gamma C^*}{\gamma} \leq \frac{CRC^*}{\gamma} \to 0 \quad .$$

But we know that in the same notation as before, $CQ_\gamma C^*$ is monotone decreasing as γ increases, and hence it follows that

$$CQ_\gamma C^* \to 0 \quad .$$

Hence in the Kalman filter:

$$\hat{s}_n = C\hat{x}_n \quad ;$$

$$\hat{x}_n = \left(I - \frac{P_\gamma C^* C}{\gamma}\right)Ax_{n-1} + \frac{P_\gamma C^*}{\gamma} v_n$$

in the limit as $\gamma \to \infty$, the system matrix

$$\left(I - \frac{P_\gamma C^* C}{\gamma}\right) A \to A \quad ,$$

while

$$\frac{P_\gamma C^*}{\gamma} \quad ,$$

of course, goes to zero so that $\hat{s}_n \to 0$. However, the normalized gain frequency function, since it depends only on the system matrix, is now the same as that of the signal. In other words, the conclusions in the examples hold also in the general case.

4.4. ON-LINE ESTIMATION OF SYSTEM PARAMETERS

In this section we shall study a generic problem of "System Identification": estimation of system parameters. We shall treat this as an application of the estimation theory developed in Chapter 3, as well as an application of Kalman filtering theory, in which the novelty is that we need to examine the asymptotic behavior of the filter, even though it is time-varying.

Problem Statement

We are given the linear system signal model:

$$\left.\begin{aligned} v_n &= Cx_n + N_n \\ x_{n+1} &= Ax_n + BU_n \end{aligned}\right\} \quad , \qquad (4.4.1)$$

in which we do not know B, but the other matrices A and C are known, as well as the (observation) white noise variance, which we take to be the Identity. Our problem is to estimate B from the observed data $\{v_n\}$, on-line rather than batch, knowing the input-sequence $\{U_n\}$. We assume that

i) A is stable

ii) $\|U_n\| \leq M_u < \infty$

iii) $\lim_{N\to\infty} \frac{1}{N} \sum_1^N \|U_n\|^2$ exists and is finite. (Input with finite power.)

In accordance with the theory developed in Chapter 3, we may consider MULE and/or MLE. We shall consider the former since we shall only be interested in asymptotically efficient estimates and we shall show that the performance of our estimates is asymptotically independent of the initial variances assumed.

To be specific, let A be $n \times n$, C be $m \times n$ and B be $n \times p$ so that the input sequence U_m is $p \times 1$. Solving (4.4.1), we obtain:

$$x_n = A^n x_0 + \sum_0^{n-1} A^j B\, U_{n-1-j} \quad . \tag{4.4.2}$$

Let $\{e_i\}$ be the coordinate unit vectors in R^p (of the form: Col. (1, 0, 0, 0, ..., 0), etc.), and let

$$U_m = \sum_1^p u_m^i e_i \quad .$$

We want now to consider U_m as a linear transformation on B rather than the other way around. For this purpose, let:

$$\tilde{U}_m = \left| u_m^1 I \quad u_m^2 I \quad \cdots \quad u_m^p I \right| , \qquad (4.4.3)$$

where I is $n \times n$, so that $\tilde{U}_m$ is $n \times np$. Then we can write

$$\sum_1^p u_m^i (Be_i) = \tilde{U}_m \begin{vmatrix} Be_1 \\ Be_2 \\ \vdots \\ Be_p \end{vmatrix} .$$

From now on, we shall represent B as an $np \times 1$ column vector

$$B = \begin{vmatrix} Be_1 \\ Be_2 \\ \vdots \\ Be_p \end{vmatrix} \qquad (4.4.4)$$

and rewrite the state equations as

$$x_{n+1} = Ax_n + \tilde{U}_m B , \qquad (4.4.4a)$$

where $\tilde{U}_m$ is specified by (4.4.3) and B has the representations (4.4.4). In particular we can rewrite (4.4.2) as

$$x_n = A^n x_0 + \sum_0^{n-1} A^j \tilde{U}_{n-1-j} B$$

$$= A^n x_0 + K_n B ,$$

where

$$K_n = \sum_0^{n-1} A^j \tilde{U}_{n-1-j} .$$

The point in doing this is to facilitate the use of the theory developed in Chapter 3. Let

$$Z = \begin{vmatrix} x_0 \\ B \end{vmatrix} ;$$

$$L_n Z = CA^n x_0 + CK_n B ,$$

so that we have:

$$v = L_n Z + N_n .$$

For the MULE we assume that Z is Gaussian with covariance matrix Λ_{tr}, our guesstimate being Λ, and independent of the noise $\{N_n\}$. Then, following Chapter 3, the MULE is given by:

$$\hat{Z}_n = E[Z \mid v_1, \ldots, v_n] \tag{4.4.5}$$

$$= (\Lambda_{tr}^{-1} + R_n)^{-1} \sum_1^n L_k^* v_k ,$$

where

$$R_n = \sum_1^n L_k^* L_k$$

$$= \begin{vmatrix} \sum_1^n A^{*k} C^* C A^k & \sum_1^n A^{*j} C^* C K_j \\ \sum_1^n K_j^* C^* C A^j & \sum_1^n K_j^* C^* C K_j \end{vmatrix} . \tag{4.4.6}$$

The estimate (4.4.5) requires "batch" processing and is furthermore burdened with having to carry along the estimate of the initial state. We shall now evolve an estimate which is recursive or "on-line," and does not involve any state estimate. For this purpose we need to assume the following "Identifiability Condition":

$$\lim_{N\to\infty} \frac{1}{N} \sum_1^N K_n^* C^* C K_n \qquad (4.4.7)$$

exists and is nonsingular.

We note that condition (4.4.7) puts additional restrictions on the input sequence. Let Λ_B denote our guesstimate for the true covariance $\Lambda_{B_{tr}}$ of B. Define the estimate

$$\hat{B}_n = (\Lambda_B^{-1} + R_n)^{-1} \sum_1^n K_m^* C^* v_m \ , \qquad (4.4.8)$$

where

$$R_n = \sum_1^n K_m^* C^* C K_m \ .$$

<u>Theorem 4.4.1</u>. Under the Identifiability Condition (and A stable) the error covariance matrix

$$E[(B - \hat{B}_n)(B - \hat{B}_n)^*] \to 0 \qquad (4.4.9)$$

as $n \to \infty$.

<u>Proof</u>. We begin by decomposing (4.4.8) as the sum of three terms

$$\hat{B}_n = (\Lambda_B^{-1} + R_n)^{-1}\left(\sum_1^n K_m^* C^* C A^m x_0\right)$$
$$+ (\Lambda_B^{-1} + R_n)^{-1}\left(\sum_1^n K_m^* C^* C K_m\right) B$$
$$+ (\Lambda_B^{-1} + R_n)^{-1}\left(\sum_1^n K_m^* C^* N_m\right) . \quad (4.4.10)$$

Let us first study the middle term; this is the term that would remain if the initial condition x_0 were zero and there were no noise as well. Hence this is the term we expect to converge to B. First of all, the middle term can be written

$$(\Lambda_B^{-1} + R_n)^{-1} R_n B \quad .$$

We need the Λ_B^{-1} term, since it is possible that R_n is singular for small n. However, R_n must be nonsingular for

$$\text{all} \quad n > \text{some} \quad N_0 \quad ,$$

by virtue of our Identifiability Condition. For, suppose we can find a subsequence R_{n_k}, $n_k \to \infty$, such that each R_{n_k} is singular. Then

$$\sum_1^{n_k} K_m^* C^* C K_m B = 0$$

for some nonzero B would imply that

$$\sum_1^{n_k} ||CK_m B||^2 = 0$$

and hence that R_m is singular for every $m \le n_k$. Hence R_n must be singular for every n, however large. Let

$$R_n B_n = 0 \quad , \qquad ||B_n|| = 1 \ .$$

A further subsequence of B_n (renumber it B_n) must converge; let B_0 denote the limit. Let R denote the limiting matrix in (4.4.7). Then

$$0 = \frac{R_n}{n} B_n \to RB_0 \qquad \text{and} \qquad ||B_0|| = 1 \ ,$$

or R is singular, which contradicts our hypothesis. Hence R_n is nonsingular for all n sufficiently large.

Let us next consider the difference

$$\begin{aligned} e_B &= B - (\Lambda_B^{-1} + R_n)^{-1} R_n B \\ &= (\Lambda_B^{-1} + R_n)^{-1} \Lambda_B^{-1} B \ . \end{aligned} \tag{4.4.11}$$

Recall that if L and M are two nonsingular self-adjoint positive definite matrices, then

$$(L+M)^{-1} \le M^{-1} \ . \tag{4.4.12}$$

To prove this, it is only necessary to observe that

$$\begin{aligned} M^{-1} - (L+M)^{-1} &= M^{-1}(L + M - M)(L + M)^{-1} \\ &= M^{-1} L (L+M)^{-1} \\ &= M^{-1}(I + ML^{-1})^{-1} \\ &= M^{-1}(M^{-1} + L^{-1})^{-1} M^{-1} \end{aligned}$$

is nonnegative definite. In particular, we have that

$$||(L+M)^{-1}||_o \leq ||M^{-1}||_o ,$$

where the subscript o indicates "operator" norm (see [2] for example). Hence from (4.4.11) it follows that

$$||e_N|| \leq ||R_n^{-1}||_o \; ||\Lambda_N^{-1}B|| .$$

We shall now show that

$$||R_n^{-1}||_o$$

goes to zero. From our Identifiability Condition where the limit R is required to be nonsingular it follows that

$$\left(\frac{R_n}{n}\right)^{-1} \rightarrow R^{-1} .$$

Or, given any arbitrarily small $\varepsilon > 0$, we can find N_ε such that for all $n > N_\varepsilon$

$$\left\| R^{-1} - \left(\frac{R_n}{n}\right)^{-1} \right\|_o < \varepsilon .$$

Hence

$$\left\| \left(\frac{R_n}{n}\right)^{-1} \right\|_o \leq ||R^{-1}||_o + \varepsilon , \qquad n \rightarrow N_\varepsilon ,$$

or

$$||R_n^{-1}||_o \leq \frac{1}{n}(\varepsilon + ||R^{-1}||_o)$$

$$\leq \frac{2}{n}||R^{-1}||_o \tag{4.4.13}$$

for <u>all</u> n sufficiently large by choosing

$$\varepsilon < ||R^{-1}||_o \quad .$$

Hence

$$||e_B|| \to 0 \tag{4.4.14}$$

for each B, as $n \to \infty$. Let us calculate now the variance of e_B, regarding B as a Gaussian random variable. We have

$$E[e_B e_B^*] = (\Lambda_B^{-1} + R_n)^{-1} \Lambda_B^{-1} \Lambda_{B_{tr}} \Lambda_B^{-1} (\Lambda_B^{-1} + R_n)^{-1} \quad ,$$

which clearly goes to zero, regardless of our not knowing what $\Lambda_{B_{tr}}$ is.

Let us next tackle the other two terms in (4.4.10). The first term therein is the response to the initial condition x_0; let us show that this goes to zero for each x_0, even though we do not know what the latter is. By virtue of (4.4.13) the first term in norm is

$$\leq \frac{2||R^{-1}||_o}{n} \left\| \sum_1^n K_m^* C^* C A^m x_0 \right\| \quad . \tag{4.4.15}$$

Now

$$||K_m|| \leq \left(\sum_0^{m-1} ||A^k||_o \right) M_u$$

$$\leq \sum_0^{\infty} ||A^k||_o \, M_u \quad ,$$

and the series

$$\sum_0^{\infty} ||A^k||_o$$

converges by virtue of our assumption of stability of A. Hence

$$\left\| \sum_{1}^{n} K_m^* C^* C A^n x_0 \right\| \leq \left(\sum_{0}^{\infty} \|A^k\|_o \right)^2 \|C^* C\| \; \|x_0\| M_u \quad ,$$

from which it follows that (4.4.15) goes to zero. It is easy to see that the variance of the error due to the response to the initial condition also goes to zero; in fact, the variance

$$= (\Lambda_B^{-1} + R_n)^{-1} \left(\sum_{1}^{n} \sum_{1}^{n} K_m^* C^* C A^m \Lambda_{x_{tr}} A^{*j} C^* C K_j \right) (\Lambda_B^{-1} + R_n)^{-1} \tag{4.4.16}$$

and by virtue of our estimate the double sum in parentheses is bounded in norm, while the outer factors go to zero.

Finally, the third term in (4.4.10) is the error due to noise. Its variance is

$$= (\Lambda_B^{-1} + R_n)^{-1} \left(\sum_{1}^{n} K_m^* C^* C K_m \right) (\Lambda_B^{-1} + R_n)^{-1}$$

$$= (\Lambda_B^{-1} + R_n)^{-1} R_n (\Lambda_B^{-1} + R_n)^{-1} \quad , \tag{4.4.17}$$

which in operator norm

$$\leq \|R_n^{-1}\|_o \; \|R_n\|_o \; \|R_n^{-1}\|_o$$

$$\leq \frac{4}{n} \left\| \frac{1}{n} R_n \right\|_o \|R^{-1}\|_o^2 \qquad \text{(for } n \text{ sufficiently large)}$$

and hence

$$\to 0 \quad \text{as} \quad n \to \infty \quad .$$

This concludes the proof of the Theorem. We do remark, however, that the convergence of the error to zero also holds "with probability one," and we have essentially proved this although a detailed statement would involve specifying the sample space (sequences, in our case); and the necessary mathematical machinery will take us too far afield for inclusion here.

On-Line Version

We may now construct our on-line version of (4.4.8) by relating $\hat{B}_n$ to $\hat{B}_{n-1}$ in the following way.

$$\hat{B}_n = (\Lambda_B^{-1} + R_n)^{-1}\left(K_n^* C^* v_n + \sum_1^{n-1} K_m^* C^* v_m\right)$$

$$= (\Lambda_B^{-1} + R_n)^{-1} K_n^* C^* v_n + (\Lambda_B^{-1} + R_n)^{-1}(\Lambda_B^{-1} + R_{n-1})\hat{B}_{n-1} ,$$

which upon substituting for R_{n-1}:

$$R_{n-1} = R_n - K_n^* C^* C K_n$$

yields:

$$\hat{B}_n = \hat{B}_{n-1} + (\Lambda_B^{-1} + R_n)^{-1} K_n^* C^* (v_n - CK_n \hat{B}_{n-1}) . \quad (4.4.18)$$

We may consolidate the on-line equations finally as follows. Let

$$P_n = (\Lambda_B^{-1} + R_n)^{-1} .$$

Then we have

$$\hat{B}_n = \hat{B}_{n-1} + P_n K_n^* C(v_n - CK_n \hat{B}_{n-1}) ;$$

$$K_n = AK_{n-1} + U_{n-1} ,$$

$$K_0 = 0 ; \quad (4.4.19)$$

$$P_n = P_{n-1} - P_n K_n^* C_n^* C K_n P_{n-1} ;$$

$$P_0 = \Lambda_B .$$

Application of Kalman Filtering

The form of our on-line equations (4.4.19) clearly bears strong resemblance to the Kalman filter equations. In fact, it is possible to cast the estimation problem in such a way that we can use the Kalman filter formalism.

The basic technique -- which is useful in a variety of similar situations -- is that of "state enhancement": of appropriately incorporating additional state variables. Thus if we add the equation:

$$B_{n+1} = B_n$$

and modify our original state equation (4.4.1) to

$$x_{n+1} = Ax_n + \tilde{U}_n B_n \quad ,$$

we still retain (4.4.1) by making the initial condition

$$B_0 = B \quad .$$

The observation equation, of course, is the same as before. Let

$$Y_m = \begin{vmatrix} x_m \\ B_m \end{vmatrix}$$

(recall that B_m has dimension $np \times 1$), so that Y_m is $(n+np) \times 1$. Then we obtain the system

$$\begin{aligned} Y_{n+1} &= A_n Y_n \quad ; \\ v_n &= CY_n + N_n \quad , \end{aligned} \tag{4.4.20}$$

where $\mathcal{A}_n$ is the compound matrix $((n+np) \times (n+np))$:

$$\mathcal{A}_n = \begin{bmatrix} A & \tilde{U}_n \\ 0 & I \end{bmatrix} ,$$

where I is the $np \times np$ unit matrix, and $\mathcal{C}$ is the compound matrix

$$\mathcal{C} = | C \quad 0 | \quad .$$

We note now that

$$\hat{Y}_n = E[Y_n \mid v_1, \ldots, v_n]$$

$$= \begin{vmatrix} \hat{x}_n \\ \hat{B}_n \end{vmatrix} ,$$

where

$$\hat{B}_n = E[B_n \mid v_1, \ldots, v_n]$$

$$= E[B \mid v_1, \ldots, v_n \;]$$

since

$$B_n = B \quad .$$

The Kalman filter equations are:

$$\hat{Y}_{n+1} = \mathcal{A}_n \hat{Y}_n + \mathcal{P}_{n+1} \mathcal{C}^* (v_{n+1} - \mathcal{C}\mathcal{A}_n \hat{Y}_N) \quad ; \qquad (4.4.21)$$

$$\mathcal{P}_{n+1} = (I - \mathcal{P}_{n+1} \mathcal{C}^* \mathcal{C}) \; \mathcal{A}_n \mathcal{C}_n \mathcal{A}_n^*$$

$$\mathcal{P}_0 = \begin{bmatrix} \Lambda_{x_{tr}} & 0 \\ 0 & \Lambda_{x_{tr}} \end{bmatrix} ;$$

$$\hat{Y}_0 = \begin{vmatrix} 0 \\ 0 \end{vmatrix} \quad .$$

To see the relationship of (4.4.19) to (4.4.21), we note first of all that

$$CA_n\hat{Y}_n = CA\hat{x}_n + \tilde{U}_n\hat{B}_n \quad ;$$

and that we can write:

$$\mathcal{P}_n = \begin{bmatrix} P_n^{11} & P_n^{12} \\ P_n^{21} & P_n^{22} \end{bmatrix} ,$$

where

$$P_n^{11} = E[(x_n - \hat{x}_n)(x_n - \hat{x}_n)^*] \quad ;$$

$$P_n^{22} = E[(B_n - \hat{B}_n)(B_n - \hat{B}_n)^*] \quad .$$

Hence

$$\hat{x}_{n+1} = A\hat{x}_n + \tilde{U}_n\hat{B}_n + P_{n+1}^{11}C^*(v_{n+1} - CA\hat{x}_n - C\tilde{U}_n\hat{B}_n); \quad (4.4.22)$$

$$\hat{B}_{n+1} = \hat{B}_n + P_{n+1}^{21}C^*(v_{n+1} - CA\hat{x}_n - C\tilde{U}_n\hat{B}_n) \quad . \quad (4.4.23)$$

Our previous estimate of B, defined by (4.4.19), is suboptimal. And since it has been shown to be asymptotically efficient, it follows that so is the optimal Kalman estimate B_n. We can also show this more directly, as follows. Solving the state equation in (4.4.21), we have:

$$Y_n = \begin{vmatrix} A^n & K_n \\ 0 & I \end{vmatrix} Y_0 \quad ,$$

where K_n is defined as before. From which, using (4.4.5) and (3.5.12), it follows that:

$$\mathcal{P}_n = \begin{bmatrix} A^n & K_n \\ 0 & I \end{bmatrix} Q_n \begin{bmatrix} A^{*n} & 0 \\ K_n^* & I \end{bmatrix} , \quad (4.4.24)$$

where

$$Q_n = (\Lambda_{tr}^{-1} + R_n)^{-1} \quad ;$$

$$\Lambda_{tr} = \begin{bmatrix} \Lambda_{x_{tr}} & 0 \\ 0 & \Lambda_{B_{tr}} \end{bmatrix}$$

and R_n is given by (4.4.6). Let

$$Q_n = \begin{bmatrix} Q_n^{11} & Q_n^{12} \\ Q_n^{21} & Q_n^{22} \end{bmatrix} ,$$

where Q_n^{11} is $n \times n$ and Q_n^{22} is $np \times np$. Then we can readily calculate that

$$P_n^{22} = Q_n^{22} \quad ; \tag{4.4.25}$$

$$P_n^{21} = Q_n^{21} A^{*n} + Q_n^{22} K_n^* \quad ; \tag{4.4.26}$$

$$P_n^{11} = A^n Q_n^{11} A^{*n} + K_n Q_n^{21} A^{*n} + A^n Q_n^{21} K_n^* + K_n Q^{22} K_n^* . \tag{4.4.27}$$

Now from (see Problem 4.1.8)

$$P_{n+1} \le A_n P_n A_n^*$$

it follows that P_n^{22} is monotone decreasing. (Of course, this is also evident from (4.4.25), since Q_n is monotone decreasing.)

Let us next calculate Q_n. Using the notation

$$\mathcal{R}_n = \begin{bmatrix} R_n^{11} & R_n^{12} \\ R_n^{21} & R_n^{22} \end{bmatrix} ,$$

where

$$R_n^{22} = \Lambda_{B_{tr}}^{-1} + \sum_1^n K_j^* C^* C K_j \quad ;$$

$$R_n^{11} = \sum_1^n A^{*k} C^* C A^k + \Lambda_{x_{tr}}^{-1} \quad ;$$

$$R_n^{12} = \sum_1^n A^{*j} C^* C K_j$$

$$= (R_n^{21})^* \quad ,$$

we have

$$Q_n^{11} = R_n^{11^{-1}} - R_n^{11^{-1}} R_n^{12} \; R_n^{21} R_n^{11^{-1}} R_n^{12} - R_n^{22} \;{}^{-1} R_n^{21} R_n^{11^{-1}} ; \quad (4.4.28)$$

$$Q_n^{22} = R_n^{22^{-1}} - R_n^{22^{-1}} R_n^{21} \left(R_n^{12} R_n^{22^{-1}} R_n^{21} - R_n^{11} \right)^{-1} R_n^{12} R_n^{22^{-1}} , \quad (4.4.29)$$

where the matrices in parentheses are nonsingular since R_n is nonsingular. We recall now our estimate (4.4.13) from which it follows that for sufficiently large n:

$$\left\| R_n^{22^{-1}} \right\|_o \leq \frac{2}{n} \| R^{-1} \|_o \quad .$$

Now both R_n^{11} and R_n^{12} converge and

$$(R_n^{11})^{-1} \to \left(\Lambda_{x_{tr}}^{-1} + \sum_1^{\infty} A^{*k} C^* C A^k \right)^{-1} = (R_\infty^{11})^{-1} .$$

Hence it follows that

$$Q_n^{22} \to 0 \quad ;$$

$$Q_n^{11} \to (R_\infty^{11})^{-1} \quad .$$

Hence also

$$Q_n^{12} \text{ and } Q_n^{21} \to 0$$

since (Schwarz inequality)

$$\|Q_n^{12}\| \leq \sqrt{\|Q_n^{11}\|}\sqrt{\|Q_n^{22}\|} \quad .$$

Finally from (4.4.25) and (4.4.27) it follows that P_n goes to zero.

Of course, the true values of the variance matrices $\Lambda_{x_{tr}}$ and $\Lambda_{B_{tr}}$ cannot be considered to be known, so that one must replace them by guesstimates Λ_x and Λ_B. In that case, our estimates are no longer optimal. Hence we define the sub-optimal filter:

$$\hat{x}_{n+1}^a = A\hat{x}_n^a + \tilde{U}_n\hat{B}_n + P_{n+1}^{a,11} C^*(v_{n+1} - CA\hat{x}_n^a - C\tilde{U}_n\hat{B}_n) \quad ;$$

$$\hat{B}_{n+1}^a = \hat{B}_n^a + P_{n+1}^{a,21} C^*(v_{n+1} - CA\hat{x}_n^a - C\tilde{U}_n\hat{B}_n) \quad , \qquad (4.4.30)$$

where

$$P_{n+1}^a = (I - P_{n+1}^a C^* C)A_n P_n^a A_n^*(I - P_{n+1}^a C^* C)^* + P_{n+1}^a C^* C P_{n+1}^a \quad ;$$

$$P_0^a = \begin{bmatrix} \Lambda_x & 0 \\ 0 & \Lambda_B \end{bmatrix} \quad ; \qquad P_n^a = \begin{bmatrix} P_n^{a,11} & P_n^{a,12} \\ P_n^{a,21} & P_n^{a,22} \end{bmatrix} \quad .$$

Let

$$T_n = E[(x_n - \hat{x}_n^a)(x_n - \hat{x}_n^a)^*] \quad .$$

Then (cf. Section 4.1) we have:

$$P_n^a - T_n = (I - P_n^a C^* C) A_{n-1} (P_{n-1}^a - T_{n-1}) A_{n-1}^* (I - C^* C P_n^a). \quad (4.4.31)$$

We know that P_n^a goes to zero, since the considerations for P_n also hold for P_n^a by replacing $\Lambda_{B_{tr}}$ by Λ_B and $\Lambda_{x_{tr}}$ by Λ_x. Moreover, from

$$P_n^a = \psi_{n-1} P_{n-1}^a \psi_{n-1}^* + P_n^a C^* C P_n^a$$

it follows that

$$\psi_n \psi_{n-1} \cdots \psi_0 P_0^a \psi_0^* \psi_1^* \cdots \psi_n^* \to 0 \quad .$$

Hence

$$[P_0^a x_n, x_n] \to 0 \quad ,$$

where

$$x_n = \psi_0^* \psi_1^* \cdots \psi_n^* x \quad .$$

Since P_0^a is nonsingular, it follows that

$$||x_n|| \to 0 \quad .$$

But from (4.4.31) we can write

$$\begin{aligned} [(P_n^a - T_n)x, x] &= [(P_0^a - T_0)x_n, x_n] \\ &\leq ||P_0^a - T^0|| \, ||x_n||^2 \\ &\to 0 \quad . \end{aligned}$$

Hence

$$||P_n^a - T_n|| \to 0 \quad .$$

Hence

$$||T_n|| \to 0 ,$$

since

$$||P^a_n|| \to 0 .$$

Remark. While (4.4.30) is more complicated than (4.4.19), the former has the advantage that it provides optimal estimates for both B_n and x_n simultaneously. We note also that the rate of convergence of $\hat{B}_n$ to B is influenced by the choice of P^a_0.

Necessary and Sufficient Conditions for Identifiability

Let us next explore the Identifiability Condition (4.4.7) and determine conditions for it to hold in terms of the system. Let us first consider a sinusoidal input, taking $p = 1$ for simplicity:

$$u^1_m = a \cos (2\pi\lambda_o m)$$

for some λ_o, and

$$0 \le \lambda_o \le \tfrac{1}{2}$$

(note that $\lambda_o = 0$ yields a "step" input). For this case we can readily verify that

$$[RB,B] = \lim_m \frac{1}{m} \sum_1^m ||CK_jB||^2$$
$$= \tfrac{1}{2}\left(a^2 \left|\left| C\left(e^{2\pi i\lambda_o} I - A\right)^{-1} B\right|\right|^2 + a^2 \left|\left| C\left(e^{-2\pi i\lambda_o} - A\right)^{-1} B\right|\right|^2\right). \tag{4.4.32}$$

If a^2 is not zero, a sufficient condition for (4.4.7) to hold is that C is nonsingular, as we can readily deduce from (4.4.32). Let

$$\psi(\lambda) = (e^{2\pi i\lambda} I - A)^{-1} .$$

Then for (4.4.32) to be zero, it is necessary that C be singular and that for some nonzero B:

$$aC\psi_r(\lambda_o)B = aC\psi_I(\lambda_o)B = 0 ,$$

where

$$\psi_r(\lambda) = \text{Re.}\ \psi(\lambda) ;$$

$$\psi_I(\lambda) = \text{Im.}\ \psi(\lambda) .$$

For the more general case, it is convenient to use the notation of (4.4.1). Thus let

$$v_m = Cx_m ;$$

$$x_{m+1} = Ax_m + \sum u_m^i B_i ,$$

where, recalling the notation in (4.4.4),

$$B_i = Be_i .$$

Then

$$[RB,B] = \lim_{n\to\infty} \frac{1}{n} \sum_1^n ||x_m||^2 ,$$

which (cf. Chapter 2)

$$= \int_{-\frac{1}{2}}^{\frac{1}{2}} \text{Tr. } p_v(\lambda)\ d\lambda ,$$

where $p_v(\lambda)$ is the spectral density of the process $\{v_m\}$ and may be calculated in terms of the spectral density $p(\lambda)$ of the process $\{U_n\}$ as

$$p_v(\lambda) = \sum_1^p \sum_1^p C(e^{2\pi i\lambda} - A)^{-1} B_i B_j^*(e^{-2\pi i\lambda} - A^*)^{-1} C^* p_{ij}(\lambda) \quad ,$$

where

$$p(\lambda) = \{p_{ij}(\lambda)\} \quad , \qquad 1 \leq i \ , \ j \leq p \ .$$

Hence

$$[RB,B] = \int_{-\frac{1}{2}}^{\frac{1}{2}} \sum_1^p \sum_1^p [C(e^{2\pi i\lambda} - A)^{-1} B_i , \ C(e^{2\pi i\lambda} - A)^{-1} B_j] p_{ij}(\lambda) \ d\lambda . \tag{4.4.33}$$

If we write

$$[RB,B] = \sum_1^p \sum_1^p [R_{ij} B_j , \ B_j] \quad ,$$

we have:

$$R_{ij} = \int_{-\frac{1}{2}}^{\frac{1}{2}} \psi(\lambda)^* \ \psi(\lambda) \ p_{ij}(\lambda) \quad d\lambda \quad .$$

If $p(\lambda)$ is diagonal, (4.4.33) reduces to

$$[RB,B] = \sum_1^p \int_{-\frac{1}{2}}^{\frac{1}{2}} || \psi(\lambda) B_i ||^2 \ p_{ii}(\lambda) \quad d\lambda \quad , \tag{4.4.34}$$

from which we can deduce generally that a sufficient condition for nonsingularity of R is that C is nonsingular, assuming of course that the input has nonzero power.

★ PROBLEM ★

Problem 4.4.1

Let it be desired to find a^N that minimizes

$$\sum_1^N \| v_n - L_n a \|^2 \gamma^{N-n}$$

for each N, where $0 < \gamma < 1$ and the sequences $\{v_n\}$, $\{L_n\}$ are given (deterministic). Show that this can be formulated as a trivial System Parameter Estimation problem. (This is the so-called Adaptive Equalizer problem in Digital Communications. See [15] for example.)

Hint: Let

$$v_n = L_n a + N_n \sqrt{\gamma^n} \quad ,$$

where $\{N_n\}$ is white noise with unit variance. Then

$$a^N = E[a \mid v_1, \ldots, v_n] \quad .$$

Let

$$\bar{v}_n = \sqrt{\gamma^n}\, v_n \quad ; \qquad \bar{L}_n = \sqrt{\gamma^n}\, L_n \quad .$$

Then

$$\bar{v}_n = \bar{L}_n a + N_n$$

and our theory applies (cf. 4.4.18) as a trivial case of a memoryless system, and we have, in particular:

$$a^N = a^{N-1} + K_N(v_n - L_N a^{N-1}) \quad .$$

Moreover, our theory yields sufficient conditions for $\{a^N\}$ to converge to a as N goes to infinity, viz:

$$\lim_{N\to\infty} \frac{1}{N} \sum_{1}^{N} \gamma^n L_n^* L_n > 0 .$$

This is an added benefit to the conscious use of a stochastic model in contrast to the "ad-hoc" least-squares minimization criterion.

4.5. (KALMAN) SMOOTHER FILTER

Let us now return to the general time-varying system equations as in Section 4.1, in the form:

$$x_{n+1} = A_n x_n + F_n N_n + B_n U_n ,$$

$$v_n = C_n x_n + G_n N_n ,$$

where A_n is nonsingular for every n,

$$F_n G_n^* = 0$$

and

$$G_n G_n^* > 0 \qquad \text{for every } n .$$

Let $v_1, v_2, \ldots, v_N$ denote the total available data set. Then

$$\hat{\hat{x}}_n = E[x_n \mid v_1, \ldots, v_N] \tag{4.5.1}$$

is called the "smoother" estimate of x_n because it allows "interpolation" or "smoothing" of the data on "both sides" of the current (n^{th}) sample. We shall now indicate a technique for calculating (4.5.1), which has the advantage of requiring less processing than the batch version. We refer to this as the Kalman smoother (equations) since it will be based on the Kalman filter estimate: a "forward" pass on the data which yields the Kalman estimate $\hat{x}_n$ and a "backward" pass which uses only the latter, and yields the "smoother" estimate.

To begin with, we note that because of the equivalence of the innovation sequence ν_n and the data v_n, we may express $\hat{\hat{x}}_n$ in terms of ν_n, and in particular exploit the "white noise" (orthogonality) feature of ν_n. Thus let

$$\hat{\hat{x}}_n = \sum_1^N A_{n,m}\nu_m + E[x_n] \quad , \qquad (4.5.2)$$

where

$$\nu_m = v_m - C_m(A_{m-1}\hat{x}_{m-1} + B_{m-1}U_{m-1}) \quad . \qquad (4.5.3)$$

To determine the "coefficients" $A_{m,n}$, we proceed as follows. First we note that

$$\begin{aligned} \hat{x}_n &= E[x_n \mid \nu_1, \ldots, \nu_n] \\ &= E[\hat{\hat{x}}_n \mid \nu_1, \ldots, \nu_n] \\ &= \sum_1^n A_{n,m}\nu_m + E[x_n] \quad . \end{aligned} \qquad (4.5.4)$$

Substituting

$$\hat{x}_n = A_{n-1}\hat{x}_{n-1} + P_n C_n^*(G_n G_n^*)^{-1}\nu_n + B_{n-1}U_{n-1}$$

into (4.5.2), we obtain:

$$\sum_1^n A_{n,m}\nu_m - \sum_1^{n-1} A_{n-1}A_{n-1,m}\nu_m = P_n C_n^*(G_n G_n^*)^{-1}\nu_n \ .$$

Since the $\{\nu_n\}$ are orthogonal, equating "like coefficients" of ν_n, we obtain

$$A_{n,n} = P_n C_n^*(G_n G_n^*)^{-1} , \tag{4.5.5}$$

$$A_{n,m} = A_{n-1}A_{n-1,m} , \quad m \le n-1. \tag{4.5.6}$$

Thus (4.5.6) provides us with $A_{n,m}$ for $m \le n$. Next let us calculate $A_{n,n+1}$. Let

$$H_n = A_n P_n A_n^* + F_n F_n^* .$$

From (4.5.2) we have:

$$\begin{aligned} A_{n,n+1} E[\nu_{n+1}\nu_{n+1}^*] &= E[x_n \nu_{n+1}^*] \\ &= E[(x_n - \hat{x}_n)\nu_{n+1}^*] \\ &= E[e_n \nu_{n+1}^*] \\ &= P_n A_n^* C_{n+1}^* . \end{aligned}$$

Since

$$E[\nu_n \nu_n^*] = C_n H_{n-1} C_n^* + G_n G_n^*$$

we obtain

$$A_{n,n+1} = P_n A_n^* C_{n+1}^*(C_{n+1} H_n C_{n+1}^* + G_{n+1}G_{n+1}^*)^{-1} . \tag{4.5.7}$$

It turns out that we can express $A_{n,n+1}$ in terms of $A_{n+1,n+1}$ analogous to (4.5.6), provided we assume that P_n is nonsingular for every n. We shall make this assumption throughout this section. We note that P_n being nonsingular is equivalent to H_n being nonsingular. Then multiplying by H_n^{-1} on both sides of the relation:

$$H_n C_{n+1}^* = P_{n+1} C_{n+1}^* (G_{n+1} G_{n+1}^*)^{-1} (G_{n+1} G_{n+1}^* + C_{n+1} H_n C_{n+1}^*) ,$$

we obtain

$$C_{n+1}^* = H_n^{-1} P_{n+1} C_{n+1}^* (G_{n+1} G_{n+1}^*)^{*-1} (G_{n+1} G_{n+1}^* + C_{n+1} H_n C_{n+1}^*).$$

Substituting this into (4.5.7), we have

$$A_{n,n+1} = P_n A_n^* H_n^{-1} P_{n+1} C_{n+1}^* (G_{n+1} G_{n+1}^*)^{-1}$$

$$= P_n A_n^* H_n^{-1} A_{n+1,n+1} , \qquad (4.5.8)$$

using (4.5.5). Let us use the notation

$$S_n = P_n A_n^* H_n^{-1} \qquad (4.5.9)$$

so that

$$A_{n,n+1} = S_n \ A_{n+1,n+1} . \qquad (4.5.9a)$$

Now

$$A_n \hat{x}_n = \sum_1^n A_n \ A_{n,m} \ \nu_m$$

$$= \sum_1^n A_{n+1,m} \ \nu_m , \qquad (4.5.10)$$

using (4.5.6). Next we can calculate that

$$(\hat{\hat{x}}_n - \hat{x}_n) = \sum_{n+1}^N A_{n,m} \ \nu_m . \qquad (4.5.11)$$

And, using (4.5.10),

$$\hat{\hat{x}}_{n+1} - A_n \hat{x}_n = \sum_{n+1}^N A_{n+1,m} \ \nu_m + B_n U_n . \qquad (4.5.12)$$

It is natural to try to express (4.5.11) in terms of (4.5.12). In fact we shall now show that

$$\hat{\hat{x}}_n - \hat{x}_n = S_n(\hat{\hat{x}}_{n+1} - A_n\hat{x}_n - B_nU_n) \quad . \quad (4.5.13)$$

We shall do this by showing that the difference is uncorrelated with ν_m for every m. First of all, this is clear for $m \leq n$. For $m = n+1$, this follows also by our construction, from (4.5.9a). Since

$$E[\hat{x}_n \nu_{n+p}^*] = 0 \qquad \text{for} \quad p \geq 1 \ ,$$

it follows that

$$E[((\hat{\hat{x}}_n - \hat{x}_n) - S_n(\hat{\hat{x}}_{n+1} - A_n\hat{x}_n - B_nU_n))\nu_{n+p}^*]$$

$$= E[(\hat{\hat{x}}_n - S_n\hat{\hat{x}}_{n+1})\nu_{n+p}^*] \ ,$$

which by the optimality of $\hat{\hat{x}}_n$, $\hat{\hat{x}}_{n+1}$

$$= E[(x_n - S_n x_{n+1})\nu_{n+p}^*] \quad . \quad (4.5.14)$$

This is zero for $p = 1$ (as we can also verify directly). Suppose it is zero for some $p \geq 1$. We shall show that it is true for $p+1$. For this purpose we note that we can write:

$$\nu_m = C_m\zeta_{m-1} + G_mN_m \ , \quad (4.5.15)$$

where

$$\zeta_m = A_m e_m + F_m N_m$$

$$e_m = x_m - \hat{x}_m \quad .$$

Now, because of the independence of x_n and N_m for $m \geq n$, it follows from (4.5.15) that

$$E[(x_n - S_n x_{n+1})\nu_m^*] = E[(x_n - S_n x_{n+1})e_{m-1}^* A_{m-1}^*] \quad (4.5.16)$$

for $m \geq n+2$. By the induction hypothesis,

$$0 = E[(x_n - S_n x_{n+1})\nu_{n+p}^*] ,$$

or

$$0 = E[(x_n - S_n x_{n+1})e_{n+p-1}^*] \quad (4.5.17)$$

by (4.5.15), since A_m^* is nonsingular. Again using (4.5.16) we have that

$$E[(x_n - S_n x_{n+1})\nu_{n+p+1}^*] = E[(x_n - S_n x_{n+1})e_{n+p}^*]A_{n+p}^* . \quad (4.5.18)$$

But by the difference equation for $\{e_n\}$ we know that e_{n+p} can be expressed in terms of e_{n+p-1}, $F_{n+p-1}N_{n+p-1}$, and N_{n+p}. The last two terms being independent of x_n and x_{n+1}, it follows that (4.5.18) must be zero by (4.5.17). Hence (4.5.14) is zero for every $p \geq 1$, and in particular it follows that

$$A_{m,n} = S_n A_{n+1,m} \quad \text{for every } m \geq n+1 . \quad (4.5.9b)$$

Thus the (backward) Kalman smoother equations are:

$$\left.\begin{aligned} \hat{\hat{x}}_n &= \hat{x}_n + S_n(\hat{\hat{x}}_{n+1} - A_n\hat{x}_n - B_n U_n) , \qquad n \leq N \\ \hat{\hat{x}}_N &= \hat{x}_N \end{aligned}\right\} . \quad (4.5.13a)$$

Let us calculate the error covariance corresponding to the smoother:

$$P_n^N = E[(x_n - \hat{\hat{x}}_n)(x_n - \hat{\hat{x}}_n)^*] .$$

We may clearly set the input U_n to be zero. Let

$$\hat{\hat{e}}_n = x_n - \hat{\hat{x}}_n .$$

Then from (4.5.13) we have by subtracting x_n from both sides:

$$\hat{\hat{x}}_n - x_n - S_n \hat{\hat{x}}_{n+1} = \hat{x}_n - x_n - S_n A_n \hat{x}_n ,$$

or

$$S_n \hat{\hat{x}}_{n+1} - \hat{\hat{e}}_n = S_n A_n \hat{x}_n - e_n .$$

On each side the variables are independent. Hence taking covariances, we have

$$S_n E[\hat{\hat{x}}_{n+1} \hat{\hat{x}}^*_{n+1}] S^*_n + P^N_n = S_n A_n E[\hat{x}_n \hat{x}^*_n] A^*_n S^*_n + P_n . \qquad (4.5.19)$$

But

$$E[\hat{\hat{x}}_{n+1} \hat{\hat{x}}^*_{n+1}] = R_{n+1} - P^N_{n+1} ,$$

$$E[\hat{x}_n \hat{x}^*_n] = R_n - P_n ,$$

where

$$R_n = E[x_n x^*_n]$$

$$= A_{n-1} R_{n-1} A^*_{n-1} + F_{n-1} F^*_{n-1} .$$

Hence substituting these in (4.5.19), we have

$$P^N_n = S_n P^N_{n+1} S^*_n - S_n R_{n+1} S^*_n + P_n$$

$$= P_n + S_n (P^N_{n+1} - A_n P_n A^*_n - F_n F^*_n) S^*_n \qquad (4.5.20)$$

and of course

$$P^N_N = P_N .$$

It is of interest to specialize to the time-invariant case (and zero input) to consider the asymptotic behavior of the smoother. For this purpose we take

$$n = \frac{N}{2}$$

and allow N to go to infinity, corresponding to the data extending symmetrically both sides in time to infinity. Let

$$P^{\infty} = \lim_{N\to\infty} P^{N}_{\frac{N}{2}} , \qquad P_{\infty} = \lim_{N\to\infty} P_n .$$

With

$$A_n = A ,$$

$$F_n = F ,$$

$$G_n = G ,$$

$$C_n = C ,$$

and (A-F) controllability, and for simplicity, A-stable, we see that

$$P^{\infty} = P_{\infty} + S_{\infty}(P^{\infty} - AP_{\infty}A^{*} - FF^{*})S_{\infty}^{*} ,$$

$$S_{\infty} = P_{\infty}A^{*}(AP_{\infty}A^{*} + FF^{*})^{-1} .$$

We can rewrite (4.5.21) as

$$\begin{aligned} P^{\infty} &= S_{\infty}P^{\infty}S_{\infty}^{*} - S_{\infty}(AP_{\infty}A^{*} + FF^{*})S_{\infty}^{*} + P_{\infty} \\ &= S_{\infty}P^{\infty}S_{\infty}^{*} - P_{\infty}A^{*}(AP_{\infty}A^{*} + FF^{*})^{-1}AP_{\infty} + P_{\infty}. \qquad (4.5.21a) \end{aligned}$$

To "solve" (4.5.21a) for P^{∞}, let us note that S_{∞} is stable. Since we are assuming that

$$H_n = AP_nA^{*} + FF^{*}$$

is nonsingular, it follows that P_n is nonsingular, and further that P_∞ is nonsingular. Moreover, taking $GG^* = \text{Identity}$ for simplicity, we know (cf. 4.2.31) that

$$(AP_\infty A^* + FF^*) = (I - P_\infty C^* C)^{-1} P_\infty$$

and hence

$$(AP_\infty A^* + FF^*)^{-1} = P_\infty^{-1}(I - P_\infty C^* C)$$

$$= (P_\infty^{-1} - C^* C) \quad ,$$

so that

$$S_\infty = P_\infty A^*(P_\infty^{-1} - C^* C)$$

$$= P_\infty(A^*(I - C^* C P_\infty))P_\infty^{-1} \quad .$$

Since $(I - P_\infty C^* C)A$ is stable, it follows that S_∞^* is stable. Hence we can solve (4.5.21a) as

$$P_\infty = \sum_0^\infty S_\infty^k (P_\infty - S_\infty A P_\infty) S_\infty^{*k} \quad .$$

Note moreover that

$$S_\infty^{*k} = P_\infty^{-1} \psi^k P_\infty \quad , \qquad k \geq 0 \quad ,$$

where

$$\psi = (I - P_\infty C^* C)A \quad ,$$

so that finally we have

$$P^\infty = P_\infty \left(\sum_0^\infty \psi^{*k} (P_\infty - S_\infty A P_\infty) \psi^k \right) P_\infty^{-1} \quad . \quad (4.5.21b)$$

We can derive an alternate formula for P^∞ in terms of the spectral density of the process $\{x_n\}$. In fact, in the steady state, the smoother estimate

$$E[x_n \mid v_m, \quad -\infty<m<\infty] = \hat{\hat{x}}_n$$

must satisfy

$$E[(x_n - \hat{\hat{x}}_n)v_m^*] = 0 \qquad \text{for every} \quad m\,. \tag{4.5.22}$$

Let

$$\hat{\hat{x}}_n = \sum_{-\infty}^{\infty} W_k v_{n-k} = \sum_{-\infty}^{\infty} W_{n-k} v_k \quad .$$

Then (4.5.22) yields:

$$E[x_n x_m^*]C^* = \sum_{-\infty}^{\infty} W_k \; E[v_{n-k} v_m^*] \quad ,$$

$$E[x_n x_m^*] = R(n-m) \quad .$$

Then (4.5.22) becomes

$$R(n-m)C^* = W_{n-m}GG^* + \sum_{-\infty}^{\infty} W_k \, C\, R(n-k-m)C^* \quad ,$$

or

$$R(n)C^* = W_n GG^* + \sum_{-\infty}^{\infty} W_k \, C\, R(n-k)C^* \; . \tag{4.5.23}$$

Let

$$p(\lambda) = \sum_{-\infty}^{\infty} e^{2\pi i\lambda} \; R(k)$$

$$= (e^{2\pi i\lambda} - A)^{-1} \; FF^*(e^{-2\pi i\lambda} - A^*)^{-1}$$

and

$$\phi(\lambda) = \sum_{-\infty}^{\infty} e^{2\pi i\lambda k} \; W_k \quad .$$

Then (4.5.23) can be expressed:

$$p(\lambda)C^* = \phi(\lambda)(GG^* + Cp(\lambda)C^*)$$

or

$$\phi(\lambda) = p(\lambda)\, C^* (GG^* + Cp(\lambda)C^*)^{-1} \quad .$$

The corresponding error matrix

$$\begin{aligned}
P^\infty &= E[(x_n - \hat{\hat{x}}_n)(x_n - \hat{\hat{x}}_n)^*] \\
&= E[x_n\hat{\hat{x}}_n^*] - E[x_n\hat{\hat{x}}_n^*] \\
&= \int_{-\frac{1}{2}}^{\frac{1}{2}} (p(\lambda) - \phi(\lambda)(CP(\lambda)C^* + GG^*)\phi(\lambda)^*)\, d\lambda \\
&= \int_{-\frac{1}{2}}^{\frac{1}{2}} (p(\lambda) - p(\lambda)C^*(GG^* + Cp(\lambda)C^*)^{-1}Cp(\lambda))\, d\lambda \ , \qquad (4.5.24)
\end{aligned}$$

where we have drawn freely from the general theory of Gaussian processes reviewed in Chapter 2. In particular, the signal-estimation error

$$\begin{aligned}
CP^\infty C^* &= \int_{-\frac{1}{2}}^{\frac{1}{2}} Cp(\lambda)\; C^*(GG^* + Cp(\lambda)C^*)^{-1}GG^*\, d\lambda \\
&= \int_{-\frac{1}{2}}^{\frac{1}{2}} ((GG^*)^{-1} + (Cp(\lambda)C^*)^{-1})^{-1}\, d\lambda \ , \qquad (4.5.25)
\end{aligned}$$

assuming that $CP(\lambda)C^*$ is nonsingular.

Example

By way of a simple illustration of the asymptotic smoothing theory, let us consider the one-dimensional example (4.3.1), (4.3.2). In this case

$$p(\lambda) = \frac{f^2}{|e^{2\pi i\lambda} - \rho|^2} ,$$

and hence (4.5.24) yields

$$p^{\infty} = \int_{-\frac{1}{2}}^{\frac{1}{2}} \frac{p(\lambda)}{1 + p(\lambda)c^2} \, d\lambda$$

$$= \int_{-\frac{1}{2}}^{\frac{1}{2}} \frac{f^2}{f^2c^2 + |e^{2\pi i\lambda} - \rho|^2} \, d\lambda \qquad (4.5.26)$$

$$= \frac{f^2}{\sqrt{(1+\rho^2+f^2c^2)^2 - 4\rho^2}} . \qquad (4.5.26a)$$

Noting that

$$S_{\infty} = \frac{\rho p_{\infty}}{f^2 + \rho^2 p_{\infty}} ,$$

(4.5.21a) yields

$$p^{\infty} = p^{\infty}\left(\frac{\rho p_{\infty}}{f^2 + \rho^2 p_{\infty}}\right)^2 - \frac{\rho^2 p_{\infty}^2}{f^2 + \rho^2 p_{\infty}} + p_{\infty} \qquad (4.5.26b)$$

which is a lot more complicated than (4.5.26a) even after simplification to

$$p^{\infty} = p_{\infty} \frac{1 - \rho^2(1 - p_{\infty}c^2)}{1 - \rho^2(1 - p_{\infty}c^2)^2} .$$

We can readily (after some algebra) verify that substituting for p_∞ yields the same answer as (4.5.26a), but we omit the details. We do see without explicit calculation that $P^\infty \approx p_\infty$ for small c (high noise) and for large c (high signal), as we expect.

★ PROBLEMS ★

Problem 4.5.1

Deduce from (4.5.20) that

$$P_{n+1}^{N} \le A_n P_n A_n^* + F_n F_n^* ,$$

using only

$$P_N \le A_N P_N A_N^* + F_N F_N^*$$

and no other optimality conditions.

Problem 4.5.2

Prove that $(P_\infty - S_\infty A P_\infty)$ is self-adjoint. Is it nonnegative definite?

Problem 4.5.3

Compare the performance of the smoother versus that of the Kalman filter for high S-N ratio and for small S-N ratio in the one-dimensional example, using (4.5.28) and Section 4.3.

4.6. KALMAN FILTER: CORRELATED SIGNAL AND NOISE

In this section we generalize the results of Section 4.1 to a class of problems where we allow the signal and noise to be correlated. Thus in our model

$$\left.\begin{aligned} v_n &= C_n x_n + N_n^o \\ x_{n+1} &= A_n x_n + U_n + N_n^s \end{aligned}\right\} , \tag{4.6.1}$$

where $\{N_n^o\}$ and $\{N_n^s\}$ are white Gaussian noise processes with

$$E[N_n^o N_n^{o*}] = G_n G_n^* \quad ;$$

$$E[N_n^s N_n^{s*}] = F_n F_n^* \quad ,$$

and may be mutually correlated:

$$E[N_n^s N_n^{o*}] = J_n \tag{4.6.2}$$

not necessarily zero, although

$$E[N_n^s N_m^{o*}] = 0 \quad , \qquad n \neq m$$

as before. Let $\bar{x}_n$ denote again the one-step predictor:

$$\bar{x}_n = E[x_n \mid v_1, \ldots, v_{n-1}] \quad .$$

We define the innovation again as

$$\begin{aligned} \nu_n &= v_n - E[v_n \mid v_{n-1}, \ldots, v_1] \\ &= v_n - C_n \bar{x}_n \quad , \end{aligned} \tag{4.6.3}$$

since

$$E[N_n^o v_{n-k}^*] = 0 \quad , \qquad k \geq 1 \quad .$$

The state innovation is defined by

$$\nu_n^s = \hat{x}_n - E[x_n \mid x_{n-1}, \ldots, x_1]$$

$$= \hat{x}_n - \bar{x}_n \quad . \tag{4.6.4}$$

Next the correlation:

$$E[\nu_n^s \nu_{n-k}^*] = E[(\hat{x}_n - \bar{x}_n)\nu_{n-k}^*]$$

$$= E[((\hat{x}_n - x_n) + (x_n - \bar{x}_n))\nu_{n-k}^*]$$

$$= 0 \qquad\qquad \text{for} \quad k \geq 1$$

by optimality of x_n and $\bar{x}_n$. Hence we must have

$$\nu_n^s = K_n \nu_n \quad , \tag{4.6.5}$$

where K_n must satisfy

$$E[\nu_n^s \nu_n^*] = K_n E[\nu_n \nu_n^*] \quad . \tag{4.6.6}$$

As before, let H_{n-1} denote the one-step prediction error:

$$H_{n-1} = E[(x_n - \bar{x}_n)(x_n - \bar{x}_n)^*] \quad .$$

Also, as before, let

$$e_n = x_n - \hat{x}_n \quad ;$$

$$P_n = E[e_n e_n^*] \quad .$$

"Writing out" (4.6.5) as:

$$\begin{aligned}\hat{x}_n - \bar{x}_n &= K_n(C_n(x_n - \bar{x}_n) + N_n^o)\\ &= (\hat{x}_n - x_n) + (x_n - \bar{x}_n) \quad ,\end{aligned}$$

we have:

$$x_n - \hat{x}_n = (I - K_nC_n)(x_n - \bar{x}_n) - K_nN_n^o \ . \tag{4.6.7}$$

Now

$$E[N_n^o(x_n - \bar{x}_n)^*] = 0 \quad ,$$

and hence, taking variances on either side of (4.6.7), we have

$$P_n = (I - K_nC_n)H_{n-1}(I - K_nC_n)^* + K_nG_nG_n^*K_n^* \ . \tag{4.6.8}$$

Proceeding as in Section 4.1, let us next calculate:

$$\begin{aligned}E[\nu_n\nu_n^*] &= (C_n(x_n - \bar{x}_n) + N_n^o)(C_n(x_n - \bar{x}_n) + N_n^o)^*\\ &= C_nH_{n-1}C_n^* + G_nG_n^* \quad ;\end{aligned}$$

$$\begin{aligned}E[\nu_n^s\nu_n^*] &= E[(\hat{x}_n - x_n + x_n - \bar{x}_n)(v_n - C_n\bar{x}_n)^*]\\ &= E[(x_n - \bar{x}_n)(v_n - C_n\bar{x}_n)^*]\\ &= E[(x_n - \bar{x}_n)(C_n(x_n - \bar{x}_n) + N_n^o)^*]\\ &= H_{n-1}C_n^* \ .\end{aligned}$$

Hence, as in Section 4.1, we can deduce from (4.6.8) that

$$P_n = (I - K_nC_n)H_{n-1} \quad ; \tag{4.6.9}$$

$$P_nC_n^* = K_n(G_nG_n^*) \ . \tag{4.6.10}$$

Let us next calculate $\bar{x}_n$. We have:

$$\begin{aligned}\bar{x}_n &= E[x_n \mid v_{n-1}, \ldots, v_1] \\ &= A_{n-1}\hat{x}_{n-1} + U_{n-1} + E[N^S_{n-1} \mid v_{n-1}, \ldots, v_1] \\ &= A_{n-1}\hat{x}_{n-1} + U_{n-1} + \hat{N}^S_{n-1} \quad . \qquad (4.6.11)\end{aligned}$$

The last term is no longer zero. In fact, using

$$E[N^S_{n-1} \mid v_{n-1}, \ldots, v_1] = E[N^S_{n-1} \mid \nu_{n-1}, \ldots, \nu_1]$$

and

$$\begin{aligned}E[N^S_{n-1}\nu^*_{n-1}] &= E[N^S_{n-1}v^*_{n-1}] \\ &= E[N^S_{n-1}N^{o*}_{n-1}] \\ &= J_{n-1} \quad ,\end{aligned}$$

$$E[N^S_{n-1}\nu^*_{n-k}] = 0 \quad , \qquad k \leq 2 \quad ,$$

we obtain

$$\hat{N}^S_{n-1} = J_{n-1}(E[\nu_{n-1}\nu^*_{n-1}])^{-1}\nu_{n-1} \quad .$$

Let

$$Q_n = J_n(E[\nu_n\nu^*_n])^{-1} \quad , \qquad (4.6.12)$$

so that we have

$$\bar{x}_n = A_{n-1}\hat{x}_{n-1} + U_{n-1} + Q_{n-1}(v_{n-1} - C_{n-1}\bar{x}_{n-1}) \quad .$$

From (4.6.5) we have:

$$\hat{x}_{n-1} = \bar{x}_{n-1} + K_{n-1}(v_{n-1} - C_{n-1}\bar{x}_{n-1}) \qquad (4.6.13)$$

and, substituting this into (4.6.13), we have the filter equations:

$$\bar{x}_n = (A_{n-1} - (A_{n-1}K_{n-1} + Q_{n-1})C_{n-1})\bar{x}_{n-1} + U_{n-1}$$

$$+ (A_{n-1}K_{n-1} + Q_{n-1})v_{n-1} \tag{4.6.14}$$

and

$$\hat{x}_n = (I - K_nC_n)\bar{x}_n + K_nv_n \quad . \tag{4.6.15}$$

We may look upon (4.6.14) as the "state equation" and (4.6.15) as the "input-state-output equation," together describing the Kalman filter. It only remains to calculate K_n and Q_n. For this purpose, we begin with H_n. Using (4.6.13), we have:

$$x_n - \bar{x}_n = A_{n-1}(x_{n-1} - \hat{x}_{n-1}) + N^S_{n-1} - \hat{N}^S_{n-1} \quad , \tag{4.6.16}$$

where, by the optimality of $\hat{x}_{n-1}$ and $\hat{N}^S_{n-1}$,

$$E[(x_{n-1} - \hat{x}_{n-1})(N^S_{n-1} - \hat{N}^S_{n-1})^*] = E[(x_{n-1} - \hat{x}_{n-1})N^{S*}_{n-1}]$$

$$= E[-\hat{x}_{n-1}N^{S*}_{n-1}]$$

$$= -K_{n-1}J^*_{n-1} \quad , \tag{4.6.17}$$

using (4.6.13). Hence, taking variances on both sides of (4.6.16), we obtain:

$$H_{n-1} = A_{n-1}P_{n-1}A^*_{n-1} + F_{n-1}F^*_{n-1} - Q_{n-1}J^*_{n-1}$$

$$- A_{n-1}K_{n-1}J^*_{n-1} - J_{n-1}K^*_{n-1}A^*_{n-1} \tag{4.6.18}$$

with

$$H_0 = A_0\Lambda A^*_0 + F_0F^*_0 \quad ; \qquad P_0 = \Lambda \quad .$$

And of course from (4.6.9) and (4.6.10):

$$K_n = P_nC^*_n(G_nG^*_n)^{-1} \quad ; \tag{4.6.19}$$

$$P_n = H_{n-1}(I + C_n^*(G_nG_n^*)^{-1}C_nH_{n-1})^{-1}$$

$$= (I + H_{n-1}C_n^*(G_nG_n^*)^{-1}C_n)^{-1}H_{n-1} \quad ; \quad (4.6.20)$$

$$(E[\nu_n\nu_n^*])^{-1} = (I - (G_nG_n^*)^{-1}C_nP_nC_n^*)(G_nG_n^*) \quad . \quad (4.6.21)$$

These equations enable us to determine P_n and Q_n iteratively. Note that using (4.6.21), we have

$$Q_n = J_n(I - (G_nG_n^*)^{-1}C_nP_nC_n^*)(G_nG_n^*)^{-1} \quad . \quad (4.6.22)$$

Steady State Theory

Let us now specialize to time-invariant systems, taking

$$A_n = A \quad ;$$

$$F_n = F \quad ;$$

$$J_n = J \quad ;$$

$$G_nG_n^* = GG^* = I$$

(for simplicity of notation) .

Our primary concern will be the steady state behavior of the filter, as in Section 4.2. As therein, we make the assumptions of Theorem 4.2.3. Further we shall assume that

$$(FF^* - JJ^*)x = 0 \rightarrow FF^*x = 0 \quad . \quad (4.6.23)$$

Let us explain the significance of this assumption. From the definition of J we have that the conditional expectation

$$E[N_n^s \mid N_n^o] = JN_n^o$$

and hence

$$E[[N_n^s - JN_n^o, x]]^2 = [(FF^* - JJ^*)x, x] .$$

Thus our assumption is equivalent to saying that no component of the state noise is exactly identical with any linear combination of the components of the "measurement" (or "observation") noise.

We shall naturally exploit the theory in Section 4.2. Specializing (4.6.9), we obtain:

$$P_n = (I - P_n C^* C) H_{n-1} , \tag{4.6.24}$$

where, using (4.6.18), we have

$$\begin{aligned} H_n &= AP_nA^* + FF^* - J(I - CP_nC^*)J^* - AP_nC^*J^* - JCP_nA^* \\ &= (A - JC)P_n(A - JC)^* + FF^* - JJ^* . \end{aligned} \tag{4.6.25}$$

To prove that P_n converges, an examination of (4.6.24) and (4.6.25) shows that we need only to prove that the conditions of Theorem 4.2.3 are satisfied with $(A - JC)$ in place of A therein and $(FF^* - JJ^*)$ in place of FF^* therein. But it is readily seen that if

$$C(A - JC)^n x = 0 \qquad \text{for every } n \geq 0 ,$$

then so is

$$CA^n x = 0 \qquad \text{for every } n \geq 0 ,$$

and hence that

$$||A^k x|| \to 0 \qquad \text{as } k \to \infty .$$

But

$$(A - JC)^k x = A^k x$$

and hence x is $(A - JC)$ stable. Similarly suppose

$$(FF^* - JJ^*)(A - JC)^{*n} x = 0 , \qquad n \geq 0 .$$

This implies by our assumption (4.6.23) that

$$(FF^*)(A - JC)^{*n} x = 0$$

and

$$(JJ^*)x = 0 .$$

Hence

$$FF^* A^{*n} x = 0 , \qquad n \geq 0 .$$

Or

$$F^* A^{*n} x = 0 , \qquad n \geq 0 .$$

Hence

$$||A^{*n} x|| \to 0 .$$

But

$$A^{*n} x = (A - JC)^{*n} x ,$$

and hence x is $(A - JC)^*$ stable. Hence it follows as in Theorem 4.2.3 that P_n converges to P_∞, say, and that P_∞ is the unique self-adjoint solution of

$$\left. \begin{aligned} P_\infty &= (I - P_\infty C^* C) H_\infty \\ H_\infty &= (A - JC) P_\infty (A - JC)^* + (FF^* - JJ^*) \end{aligned} \right\} . \qquad (4.6.26)$$

Moreover, we have that

$$(A - JC)(I - P_\infty C^* C)$$

is stable. Finally, we note that we may instrument the asymptotic version as

$$\left.\begin{aligned} \bar{x}_n^a &= (I - P_\infty C^* C)(A - JC)\bar{x}_{n-1}^a + U_{n-1} \\ &\qquad + ((A - JC)P_\infty C^* + J)v_{n-1} \\ \hat{x}_n^a &= (I - P_\infty C^* C)\bar{x}_n^a + P_\infty C^* v_n \end{aligned}\right\} , \qquad (4.6.27)$$

which will be asymptotically optimal.

<u>Example</u>

Consider the one-dimensional case as in Section 4.2. Then to satisfy (4.6.23) we only need that

$$f^2 - j^2 > 0 ,$$

and we can calculate that:

$$\begin{aligned} P_\infty &= \frac{H_\infty}{1 + H_\infty c^2} \\ &= \frac{(\rho - jc)^2 P_\infty + f^2 - j^2}{1 + c^2(f^2 - j^2) + c^2(\rho - jc)^2 P_\infty} , \end{aligned}$$

or

$$c^2(\rho - jc)^2 P_\infty^2 + (1 + c^2(f^2 - j^2) - (\rho - jc)^2)P_\infty - (f^2 - j^2) = 0 .$$

Thus in (4.2.2e) we need only replace f^2 by $(f^2 - j^2)$ and ρ by $(\rho - jc)$.

★ PROBLEMS ★

Problem 4.6.1

Show that $K_{n-1}J^*_{n-1} = H_{n-1}C^*_{n-1}Q^*_{n-1}$.

Problem 4.6.2

Suppose one designs the optimal Kalman filter assuming $J = 0$. Calculate the degradation in performance specializing to the one-dimensional case in steady state.

4.7. KALMAN FILTER FOR COLORED (OBSERVATION) NOISE

So far we have assumed that the observation noise

$$N^O_n = G_n N_n$$

was white. In this section we extend our results to a class of non-white (Gaussian) noise processes. Thus we consider the model:

$$v_n = C_n x_n + H^O_n , \tag{4.7.1}$$

where

$$N^O_n = Q_{n-1}N^O_{n-1} + \eta_{n-1} \tag{4.7.2}$$

and

$$x_n = A_{n-1}x_{n-1} + U_{n-1} + N^S_{n-1} , \tag{4.7.3}$$

where $\{\eta_n\}$ is white Gaussian with covariance Λ_n, and $\{N^S_n\}$ is white Gaussian with variance $F_n F^*_n$, and $\{N^S_n\}$ is independent of $\{\eta_n\}$.

Note that we are modelling the noise process as a Gaussian Markov process. In the stationary case, where Q_n is independent of n, so that we have

$$N_n^o = QN_{n-1}^o + \eta_{n-1} ,$$

the spectral density of the noise N_n^o is

$$(e^{2\pi i\lambda} - Q)^{-1} E[\eta_n\eta_n^*](e^{-2\pi i\lambda} - Q^*)^{-1}$$

and approaches white noise as $||Q||$ goes to zero. Thus we may consider the white noise model as the special case where Q or (Q_n) is zero.

The technique is to keep "differencing" (4.7.1) until we get white noise. Thus we multiply both sides of

$$v_{n-1} = C_{n-1}x_{n-1} + N_{n-1}^o$$

by Q_{n-1} and subtract from (4.7.1), to obtain:

$$\begin{aligned} v_n - Q_{n-1}v_{n-1} &= C_nx_n - Q_{n-1}C_{n-1}x_{n-1} + N_n - Q_{n-1}N_{n-1} \\ &= C_nx_n - Q_{n-1}C_{n-1}x_{n-1} + \eta_{n-1} , \end{aligned}$$

which, using (4.7.1) and substituting for x_n, yields

$$= (C_nA_{n-1} - Q_{n-1}C_{n-1})x_{n-1} + C_nU_{n-1}C_nN_{n-1} + \eta_{n-1} . \quad (4.7.4)$$

Let

$$w_n = v_n - Q_{n-1}v_n ,$$

$$\tilde{w}_n = w_n - C_nU_{n-1} ,$$

$$y_n = x_{n-1} ,$$

$$\bar{N}_s^o = C_nN_{n-1}^s + \eta_{n-1} ,$$

$$\bar{N}_n^s = N_{n-1}^s .$$

We can then write:

$$\left.\begin{aligned} \tilde{w}_n &= \bar{C}_n y_n + \bar{N}_n^o \\ y_{n+1} &= \bar{A}_n y_n + \bar{N}_n^s + U_n \end{aligned}\right\} , \qquad (4.7.5)$$

where

$$\bar{C}_n = (C_n A_{n-1} - Q_{n-1} C_{n-1}) \quad ;$$

$$\bar{A}_n = A_{n-1} \quad ;$$

$$\bar{U}_n = U_{n-1} \quad ;$$

$$E[\bar{N}_n^o \bar{N}_n^{o*}] = C_n F_{n-1} F_{n-1}^* C_n^* + \Lambda_{n-1} \quad ;$$

$$E[\bar{N}_n^s \bar{N}_n^{o*}] = F_{n-1} F_{n-1}^* C_{n-1}^* \quad ;$$

$$E[\bar{N}_n^s \bar{N}_{n+k}^{o*}] = 0 \quad , \qquad k \geq 1 \quad k \leq -1 \quad ;$$

$$E[\bar{N}_n^s \bar{N}_n^{s*}] = F_{n-1} F_{n-1}^* \quad .$$

Hence the signal and noise are correlated, and in the notation of Section 4.6, we have

$$J_n = F_{n-1} F_{n-1} C_n^* \quad .$$

We can therefore apply the results of Section 4.6. In doing so we note that the transformation from $\{v_n\}$ into $\{\tilde{w}_n\}$:

$$\tilde{w}_n = w_n - C_n U_{n-1} \quad ;$$

$$w_n = v_n - Q_{n-1} v_{n-1} \quad ;$$

$$w_1 = v_1$$

is 1:1. In fact we can solve for $\{v_n\}$ as follows:

$$w_n = \tilde{w}_n + C_n U_{n-1} \quad ;$$

$$v_n = w_n + Q_{n-1} w_{n-1} + Q_{n-1} Q_{n-2} w_{n-2} + \cdots + Q_{n-1} Q_{n-2} \cdots Q_1 w_1 .$$

Now we need

$$\hat{x}_n = E[x_n \mid v_1, \ldots, v_n] \quad ,$$

which is clearly

$$= E[x_n \mid \tilde{w}_1, \ldots, \tilde{w}_n]$$

$$= E[y_{n+1} \mid \tilde{w}_1, \ldots, \tilde{w}_n]$$

$$= \bar{y}_{n+1} \quad . \tag{4.7.6}$$

Hence we may apply the theory of Section 4.6 to obtain the one-step predictor $\bar{y}_n$. The relevant formula is (4.6.14), and to calculate the quantities therein, we need to note the correspondence:

$$G_n G_n^* = C_n F_{n-1} F_{n-1}^* C_n^* + \Lambda_{n-1} \quad .$$

Hence we have:

$$\hat{x}_n = \bar{y}_{n+1}$$

$$= (A_{n-1} - L_{n-1} C_n) \hat{x}_{n-1} + U_{n-1}$$

$$+ L_{n-1}(v_n - Q_{n-1} v_{n-1} - C_n U_{n-1}) \quad , \tag{4.7.7}$$

where

$$L_{n-1} = A_{n-1} P_n \bar{C}_n^* T_n + F_{n-1} F_{n-1}^* C_n^* (I - T_n \bar{C}_n P_n \bar{C}_n^*) T_n ; \tag{4.7.8}$$

$$T_n = (C_n F_{n-1} F_{n-1} C_n + \Lambda_{n-1})^{-1}$$

$$P_n = (I + H_{n-1} \bar{C}_n^* T_n \bar{C}_n)^{-1} H_{n-1} \quad ; \tag{4.7.9}$$

$$H_n = (A_{n-1} - F_{n-1}F_{n-1}^* C_n^* T_n \bar{C}_n^*) P_n (A_{n-1} - F_{n-1}F_{n-1}^* C_n^* T_n \bar{C}_n)$$
$$+ F_{n-1}F_{n-1}^* - F_{n-1}F_{n-1}^* C_n^* T_n C_n F_{n-1}F_{n-1}^* \;; \quad (4.7.9a)$$

$$\bar{C}_n = C_n A_{n-1} - Q_{n-1} C_{n-1} \;.$$

Remark. Upon setting Q_n to be zero for every n, it is not difficult to see that we revert to our previous formulas in Section 4.1.

Time-Invariant Systems: Steady State Theory

Let us now specialize to time-invariant systems and examine the steady state behavior in particular. Thus we set

$$A_n = A \;;$$
$$C_n = C \;, \qquad \bar{C}_n = \bar{C} = CA - QC \;;$$
$$F_n = F \;;$$
$$\Lambda_n = I \;;$$
$$Q_n = Q \;.$$

This yields for (4.7.7):

$$\begin{aligned}\hat{x}_n &= (A - FF^*\bar{C}^*T\bar{C})(I - P_n\bar{C}^*T\bar{C})x_{n-1} + U_{n-1} \\ &\quad + ((A - FF^*C^*\bar{C})P_n\bar{C}^*T + FF^*C^*T)(v_n - Qv_{n-1} - CU_{n-1}) \ ;\end{aligned} \tag{4.7.10}$$

$$T = (CFF^*C^* + I)^{-1} \ ;$$

$$P_n = (I + H_{n-1}\bar{C}^*T\bar{C})H_{n-1} \ ;$$

$$H_n = (A - FF^*C^*T\bar{C})P_n(A - FF^*C^*T\bar{C})^* + FF^* - FF^*C^*TCFF^* \ ;$$

$$P_0 = E[x_0x_0^*] \ , \qquad E[x_0] = 0 \ ;$$

$$\hat{x}_0 = 0 \ .$$

For the convergence of P_n, we may paraphrase the conditions of Section 4.6, for the one-step predictor. Thus we need the conditions of Theorem 4.2.1 that

A^* - unstable states are F^* observable

and

A-unstable states are $\sqrt{T}(CA - QC)$ observable.

In addition we need that (corresponding to (4.6.23)):

$$(FF^* - FF^*C^*(I + CFF^*C^*)^{-1}CFF^*)x = 0 \tag{4.7.11}$$

implies that $FF^*x = 0$. Now since T is nonsingular, the second condition can be replaced by:

A-unstable states are $(CA - QC)$ observable . (4.7.12)

The condition (4.7.11) is automatically satisfied -- or is no condition at all, as is readily shown. Thus, suppose

$$C(FF^* - FF^*C^*(I + FF^*C^*)^{-1}CFF^*)x = 0 ,$$

or

$$(I - CFF^*C^*(I + CFF^*C^*)^{-1})CFF^*x = 0 ,$$

or

$$(I + CFF^*C^*)^{-1}CFF^*x = 0 .$$

Then

$$CFF^*x = 0 .$$

Hence

$$(FF^* - FF^*C^*(I + CFF^*C^*)^{-1}CFF^*)x = FF^*x = 0 .$$

★ PROBLEM ★

Problem 4.7.1

Calculate P_∞ in the one-dimensional case. Suppose one designs the optimal Kalman filter assuming $Q = 0$. Calculate the degradation in performance in the one-dimensional steady state case.

4.8. EXAMPLE

To help illustrate some of the theory and techniques we have discussed so far, we present a worked-out example based on a (gross) simplification of a problem arising in Inertial Navigation Systems (correcting for vertical deflection of gravity [4]). Thus let

$$v_n = s_n + \varepsilon N_n^3 , \tag{4.8.1}$$

$$s_n = \xi_n + \eta_n , \tag{4.8.2}$$

$$\xi_{n+1} = \rho \xi_n + \sigma N_n^1 , \tag{4.8.3}$$

$$\eta_{n+1} = \rho \eta_n + \sigma N_n^2 , \tag{4.8.4}$$

where

$$\begin{vmatrix} N_n^1 \\ N_n^2 \\ N_n^3 \end{vmatrix} \sim \text{white Gaussian with unit variance}$$

$$0 < \rho \leq 1 , \quad 0 < \sigma, \varepsilon .$$

We can rewrite this in our notation as:

$$\left.\begin{aligned} v_n &= Cx_n + GN_n \\ x_{n+1} &= Ax_n + FN_n \end{aligned}\right\} , \tag{4.8.5}$$

where

$$x_n = \begin{vmatrix} \xi_n \\ \eta_n \end{vmatrix} ,$$

$$A = \begin{bmatrix} \rho & 0 \\ 0 & \rho \end{bmatrix} , \tag{4.8.6}$$

$$F = \begin{bmatrix} \sigma & 0 & 0 \\ 0 & \sigma & 0 \end{bmatrix} , \tag{4.8.7}$$

$$C = [\ 1 \quad 1\ ] , \tag{4.8.8}$$

$$G = [\ 0 \quad 0 \quad \varepsilon\ ] . \tag{4.8.9}$$

We have

$$GG^* = \varepsilon^2 ,$$

$$FF^* = \begin{bmatrix} \sigma^2 & 0 \\ 0 & \sigma^2 \end{bmatrix} ,$$

$$C^*C = \begin{bmatrix} 1 & 1 \\ 1 & 1 \end{bmatrix} .$$

Let us first consider the structure of the Kalman filter. Let

$$P_n = E[(x_n - \hat{x}_n)(x_n - \hat{x}_n)^*] .$$

We shall assume $\rho < 1$ (and consider $\rho = 1$ as a limiting case later) so that A is stable. We note that $(C \sim A)$ is not observable but $(A \sim F)$ is controllable. Also

$$\Phi(P) = \left(I + H(P)\frac{C^*C}{\varepsilon^2}\right)^{-1} H(P) ,$$

where

$$H(P) = \sigma^2 P + \sigma^2 I ,$$

which is clearly nonsingular, so that $\Phi(P)$ is also nonsingular and

$$\Phi(P) = \left((\sigma^2 I + \sigma^2 P)^{-1} + \frac{C^*C}{\varepsilon^2}\right)^{-1} . \tag{4.8.10}$$

The SSRE is therefore

$$P_s^{-1} = (\sigma^2 I + \sigma^2 P_s)^{-1} + \frac{C^* C}{\varepsilon^2} \quad . \quad (4.8.11)$$

We can solve this by indirect methods. First of all,

$$R_n = E[x_n x_n^*]$$

satisfies

$$R_{n+1} = \sigma^2 R_n + \sigma^2 I \quad .$$

Hence if R_n is diagonal, so is R_{n+1}. We shall take R_0 to be diagonal. (Of course $R_0 = P_0$, so that P_0 is also diagonal.) In particular, therefore, ξ_n and η_n are independent for every $n \geq 0$. Also, since

$$E(x_n x_{n+m}^*) = \rho^m R_n \, , \qquad m > 0 \, ,$$

it follows that

$$E(\xi_n \eta_m) = 0 \qquad \text{for all} \quad n, m \quad .$$

In particular, therefore,

$$E[(\xi_n - \eta_n)(\xi_k + \eta_k)] = | 1 \; -1 | \, \rho^{n-k} R_k \begin{vmatrix} 1 \\ 1 \end{vmatrix} \, , \qquad k \leq n \, .$$

This is zero if

$$R_0 = \gamma I \quad ,$$

or, in other words, if

$$E[\xi_0^2] = E[\eta_0^2] \quad .$$

Let us assume this. (This is certainly true for the special case $R_0 = 0$.) Hence it follows

$$E[\xi_n - \eta_n \mid v_1, \ldots, v_n] = 0$$

or

$$\hat{\xi}_n = \hat{\eta}_n .$$

Now

$$\hat{x}_n = A\hat{x}_{n-1} + \frac{P_n C^*}{\varepsilon^2}(v_n - CA\hat{x}_{n-1}) ,$$

where

$$CA\hat{x}_{n-1} = \rho\hat{s}_{n-1} .$$

Since

$$|1 \quad -1| \ \hat{x}_n = 0 ,$$

and

$$A\hat{x}_{n-1} = \rho\hat{x}_{n-1} ,$$

we obtain that

$$|1 \quad -1| \ P_n C^* = 0$$

or

$$|1 \quad -1| \ P_n \begin{vmatrix} 1 \\ 1 \end{vmatrix} = 0 ,$$

or we can write:

$$P_n C^* = \begin{vmatrix} q_n \\ q_n \end{vmatrix} .$$

Hence we have:

$$\hat{\xi}_n = \rho\hat{\xi}_{n-1} + \frac{q_n}{\varepsilon^2}(v_n - \rho\hat{s}_{n-1}) , \qquad (4.8.12)$$

$$\hat{\eta}_n = \rho\hat{\eta}_{n-1} + \frac{q_n}{\varepsilon^2}(v_n - \rho\hat{s}_{n-1}) . \qquad (4.8.13)$$

We can also get an interpretation of q_n by noting that

$$\begin{aligned} \hat{s}_n &= C\hat{x}_n \\ &= CA\hat{x}_{n-1} + \frac{CP_n C^*}{\varepsilon^2}(v_n - \rho\hat{s}_n) , \end{aligned}$$

or

$$\hat{s}_n = \rho\hat{s}_{n-1} + \frac{CP_nC^*}{\varepsilon^2}(v_n - \rho\hat{s}_n) \quad . \tag{4.8.14}$$

Hence

$$CP_nC^* = 2q_n \quad .$$

Since $\{s_n\}$ satisfies

$$s_{n+1} = \rho s_n + \sqrt{2}\,\sigma\,\frac{(N_n^1+N_n^2)}{\sqrt{2}} \tag{4.8.14a}$$

and

$$v_n = s_n + \varepsilon N_n^3$$

we see that, letting

$$t_n = E[(s_n - \hat{s}_n)^2] \quad ,$$

we have:

$$t_{n+1} = \frac{\rho^2 t_n + 2\sigma^2}{1 + \dfrac{(\rho^2 t_n + 2\sigma^2)}{\varepsilon^2}} \quad . \tag{4.8.15}$$

And of course

$$q_n = \frac{t_n}{2} \quad . \tag{4.8.16}$$

The steady state value of t_n is therefore, denoting it by t_s, given by

$$\frac{t_s}{\varepsilon^2} = \frac{\sqrt{(1+2\lambda^2-\rho^2)^2 + 8\lambda^2\rho^2} - (1+2\lambda^2-\rho^2)}{2\rho^2} \quad , \tag{4.8.17}$$

where

$$\lambda = \frac{\sigma}{\varepsilon} \quad ;$$

and the limiting value of q_n is

$$q_s = \frac{t_s}{2} .$$

To calculate P_s from this, we first note that

$$E[(\hat{\xi}_n - \hat{\eta}_n)^2] = 0 .$$

Hence

$$2\hat{r}_n = 2E[\hat{\xi}_n \hat{\eta}_n] ,$$

where

$$\hat{r}_n = E[(\hat{\xi}_n)^2] = E[(\hat{\eta}_n)^2] .$$

From

$$t_n = E[(\xi_n - \eta_n)^2] - E[(\hat{\xi}_n + \hat{\eta}_n)^2]$$

$$= 2r_n - 4\hat{r}_n$$

we obtain:

$$\hat{r}_n = \frac{2r_n - t_n}{4} ,$$

$$= \frac{r_n - q_n}{2} .$$

Now

$$P_n = E[x_n x_n^*] - E[\hat{x}_n \hat{x}_n^*]$$

and

$$E[x_n x_n^*] = \begin{bmatrix} r_n & 0 \\ 0 & r_n \end{bmatrix} ,$$

where

$$r_{n+1} = \rho^2 r_n + \sigma^2 ,$$

so that

$$r_n = \rho^n r_0 + \sum_0^{n-1} \sigma^2 \rho^{2k} ,$$

and

$$E[\hat{x}_n \hat{x}_n^*] = \begin{bmatrix} \hat{r}_n & \hat{r}_n \\ \hat{r}_n & \hat{r}_n \end{bmatrix} .$$

Hence

$$P_n = \begin{bmatrix} r_n - \hat{r}_n & -\hat{r}_n \\ -\hat{r}_n & r_n - \hat{r}_n \end{bmatrix}$$

$$= \begin{bmatrix} \frac{q_n + r_n}{2} & \frac{q_n - r_n}{2} \\ \frac{q_n - r_n}{2} & \frac{q_n + r_n}{2} \end{bmatrix} . \tag{4.8.18}$$

Hence the steady state value

$$P_s = \lim_{n \to \infty} P_n$$

$$= \begin{bmatrix} \frac{q_s + r_s}{2} & \frac{q_s - r_s}{2} \\ \frac{q_s - r_s}{2} & \frac{q_s + r_s}{2} \end{bmatrix} , \tag{4.8.19}$$

where

$$r_s = \frac{\sigma^2}{1 - \rho^2} .$$

Or we have found the solution of the SSRE (4.8.11) which is of course the same whatever the initial P_0.

Let us consider some limiting cases. Let $\rho \to 1$. Then t_n and hence q_n converge while r_n increases without bound. Hence from (4.8.18) we see that

$$\text{Tr. } P_n = q_n + r_n \tag{4.8.20}$$

increases without bound. This is consistent with our theory, since $(C \sim A)$ is not observable, and for $\rho = 1$, A becomes unstable. Let $\rho \to 0$. In that case both ξ_n and η_n become white noises and

$$\hat{\xi}_n = \hat{\eta}_n = \tfrac{1}{2} v_n \ , \tag{4.8.21}$$

as it should be. Finally let $\varepsilon \to 0$. The observation noise goes to zero, but the Kalman filter is still well defined. In fact for this case:

$$\hat{\xi}_n = \rho\hat{\xi}_{n-1} + \tfrac{1}{2}[v_n - \rho\hat{\gamma}_{n-1}] \ ,$$

$$\hat{\eta}_n = \rho\hat{\eta}_{n-1} + \tfrac{1}{2}[v_n - \rho\hat{\gamma}_{n-1}] \ .$$

Kalman Smoother

Next let us look at the smoother. We follow the notation of Section 4.5. Our basic recursion formula is (4.5.13a) where we note that

$$S_n = P_n A^*(AP_n A^* + FF^*)^{-1}$$

$$= P_n A^*(P_n^{-1} - C^*(GG^*)^{-1}C) \ .$$

Specializing to our case,

$$S_n = \rho P_n\left(P_n^{-1} - \frac{C^*C}{\varepsilon^2}\right) = \rho\left(I - \frac{P_n C^* C}{\varepsilon^2}\right)$$

$$= \rho \begin{vmatrix} 1-\bar{q}_n & -\bar{q}_n \\ -\bar{q}_n & 1-\bar{q}_n \end{vmatrix} \ ,$$

where

$$\bar{q}_n = \frac{q_n}{\varepsilon^2} .$$

In particular we observe that S_n is self-adjoint and nonnegative definite. Hence (4.5.13a) becomes

$$\hat{\hat{\xi}}_n = \rho\hat{\hat{\xi}}_{n+1} - \rho\bar{q}_n\hat{\hat{s}}_{n+1} + (1-\rho^2)\hat{\xi}_n + \rho^2\bar{q}_n\hat{s}_n , \qquad (4.8.22)$$

$$\hat{\hat{\eta}}_n = \rho\hat{\hat{\eta}}_{n+1} - \rho\bar{q}_n\hat{\hat{s}}_{n+1} + (1-\rho^2)\hat{\eta}_n + \rho^2\bar{q}_n\hat{s}_n , \qquad (4.8.23)$$

where

$$\hat{\hat{s}}_n = \hat{\hat{\xi}}_n + \hat{\hat{\eta}}_n ,$$

and satisfies

$$\hat{\hat{s}}_n = \hat{s}_n + \rho(1 - 2\bar{q}_n)(\hat{\hat{s}}_{n+1} - \rho\hat{s}_n) ,$$

as can be verified independently from (4.5.13a) or by adding (4.8.22) and (4.8.23). Similarly, subtracting (4.8.23) from (4.8.22), it follows that

$$E[\hat{\hat{\xi}}_n - \hat{\hat{\eta}}_n]^2 = 0$$

and hence that

$$t_n^N = 2r_n - 4\hat{\hat{r}}_n ,$$

where

$$t_n^N = E[(s_n - \hat{\hat{s}}_n)^2] ,$$

$$\hat{\hat{r}}_n = E[(\hat{\hat{\xi}}_n)^2] = E[(\hat{\hat{\eta}}_n)^2] .$$

Hence

$$\hat{\hat{r}}_n = \frac{r_n - q_n^N}{2} ,$$

where

$$2q_n^N = t_n^N .$$

Hence we can write P_n^N in the form:

$$P_n^N = \begin{bmatrix} \frac{r_n + q_n^N}{2} & \frac{r_n - q_n^N}{2} \\ \frac{r_n - q_n^N}{2} & \frac{r_n + q_n^N}{2} \end{bmatrix}$$

To complete the calculation of P_n^N, we only need to note that specializing (4.5.20) to (4.8.14a), we have that

$$t_n^N = t_n + \rho^2(1 - 2\bar{q}_n)^2(t_{n+1}^N - 2\rho^2 q_n - 2\sigma^2) .$$

We can proceed to the steady state as in Section 4.5. Defining

$$t^\infty = \lim_N t_{\frac{N}{2}}^N , \qquad P^\infty = \lim_N P_{\frac{N}{2}}^N ,$$

we have that

$$t^\infty = \frac{t_s - \rho^2\left(1 - \frac{2q_s}{\varepsilon^2}\right)^2 (2\rho^2 q_s - 2\sigma^2)}{1 - \rho^2\left(1 - \frac{2q_s}{\varepsilon^2}\right)^2} ,$$

$$P^\infty = \begin{bmatrix} \frac{r_s}{2} + \frac{t^\infty}{4} & \frac{r_s}{2} - \frac{t^\infty}{4} \\ \frac{r_s}{2} - \frac{t^\infty}{4} & \frac{r_s}{2} + \frac{t^\infty}{4} \end{bmatrix}$$

The steady state matrix P^∞ can be obtained using the spectral formula (4.5.23) which in the present case yields:

$$P^{\infty} = \int_{-\frac{1}{2}}^{\frac{1}{2}} \left(p(\lambda)I - \frac{p(\lambda)^2}{\varepsilon^2 + 2p(\lambda)} C^*C \right) d\lambda \quad ,$$

$$t^{\infty} = \int_{-\frac{1}{2}}^{\frac{1}{2}} \left(2p(\lambda) - \frac{4p(\lambda)^2}{\varepsilon^2 + 2p(\lambda)} \right) d\lambda \quad ,$$

where

$$p(\lambda) = \frac{\sigma^2}{|e^{2\pi i\lambda} - \rho|^2} \quad .$$

Of course

$$\int_{-\frac{1}{2}}^{\frac{1}{2}} p(\lambda)\, d\lambda = r_s = \frac{\sigma^2}{1 - \rho^2} \quad ,$$

and we can go "backwards" and calculate the integral (or verify!)

$$\int_{-\frac{1}{2}}^{\frac{1}{2}} \frac{p(\lambda)^2}{\varepsilon^2 + 2p(\lambda)}\, d\lambda$$

as

$$t^{\infty} - 2r_s \quad .$$

Chapter 5.

LIKELIHOOD RATIOS: GAUSSIAN SIGNALS IN GAUSSIAN NOISE

In Chapter 3, we have seen the importance of the likelihood functional for estimation problems. In this chapter we shall derive likelihood ratio formulas for Gaussian signals in Gaussian noise in which Kalman theory plays an essential role.

Let us consider the general time-varying model:

$$\left.\begin{aligned} v_n &= C_n x_n + N_n^o \\ x_{n+1} &= A_n x_n + F_n N_n + B_n U_n \end{aligned}\right\} , \qquad (5.1)$$

where, as usual, $\{N_n^o\}$ is white Gaussian with unit variance matrix

$$F_n G_n^* = 0 \quad ;$$

$$E[N_n^o N_n^{o*}] = G_n G_n^* \quad \text{nonsingular} \quad ,$$

U_n is the (deterministic) input.

We want to calculate the "Likelihood Ratio" which we shall define to be the ratio:

$$\frac{\text{Joint density of } v_1,\ldots,v_n \text{ (with signal present)}}{\text{Joint density of } v_1,\ldots,v_n \text{ (with signal absent)}} . \quad (5.2)$$

To calculate the density in the numerator we need to know the mean and covariance. Let

$$v = \begin{vmatrix} v_1 \\ \vdots \\ v_n \end{vmatrix}$$

and let R denote the covariance matrix of v. Let

$$E[v_i] = m_i \quad ;$$

$$m = \begin{vmatrix} m_1 \\ \vdots \\ m_n \end{vmatrix} .$$

Then we will need to calculate

$$[R^{-1}(v-m), (v-m)] \quad . \qquad (5.3)$$

Thus for each n we will need to invert the matrix R. This inversion is most efficiently accomplished by factorization. Thus we seek to find L such that

$$R^{-1} = L^* L$$

and L is "lower triangular." But for the system (5.1) this is precisely what Kalman filtering does for us. In fact, recall that the innovation

$$\nu_n = v_n - C_n(A_{n-1}\hat{x}_{n-1} + B_{n-1}U_{n-1}) \qquad (5.4)$$

is white Gaussian with covariance

$$J_n = E[\nu_n \nu_n^*] = G_n G_n^* + C_n H_{n-1} C_n^* \;;$$

$$H_n = A_n P_n A_n^* + F_n F_n^* \;.$$

Since the transformation (5.4) is 1:1 with unit Jacobian,

$$p(v_1, \ldots, v_n) = p_1(\nu_1) \cdots p_k(\nu_k)$$

$$= \prod_{k=1}^{n} p_k(v_k - C_k(A_{k-1}\hat{x}_{k-1} + B_{k-1}U_{k-1})) \;,$$

where $p_k(\cdot)$ is the Gaussian density with mean zero and variance matrix J_k. Since the denominator in (5.2) is the joint density of the observation noise sequence with variance matrix $G_n G_n^*$, it follows that the ratio (5.2) can be expressed:

$$L(v_1, \ldots, v_n)$$

$$= \exp -\frac{1}{2}\left\{\sum_1^n [J_k^{-1}(v_k - C_k(A_{k-1}\hat{x}_{k-1}+B_{k-1}U_{k-1})), \; v_k - C_k(A_{k-1}\hat{x}_{k-1} + B_{k-1}U_{k-1})] - \sum_1^n [(G_k G_k^*)^{-1} v_k, v_k] + \sum_1^n \log |J_k| - \sum_1^n \log |G_k G_k^*| \right\} , \qquad (5.5)$$

where $|\cdot|$ denotes determinant (of a square matrix). Of course, we can also use the one-step predictor:

$$\bar{x}_k = A_{k-1}\hat{x}_{k-1} + B_{k-1}U_{k-1}$$

in (5.5) along with the one-step predictor equations (4.1.39). Formula (5.5) comprises thus the basic first step in considering maximum likelihood estimates for unknown parameters in the system (5.1). We note that to calculate the Kalman filter estimate $\hat{x}_k$, we need to know the initial covariance P_0. Since we can only guesstimate this, (5.5) is not totally accurate. Of course, the lack of knowledge of

$$E[x_0 x_0^*]$$

also means we cannot calculate R in (5.3) either. On the other hand, at least for time-invariant systems we know that under appropriate conditions the estimates $\hat{x}_n$ will be asymptotically optimal, and in fact this is virtually the only case in which we can <u>prove</u> that the estimate obtained by maximizing (5.5) is asymptotically efficient.

<u>Example</u>

In Section 4.4 we considered the problem of estimating B in the time-invariant version of (5.1) in which, moreover, there was no state noise as well ($FN_n = 0$). We can use (5.5) for this purpose. Placing the same conditions as in (4.4.1), we have:

$-2 \log L(v_1, \ldots, v_n)$

$$= \sum_1^n [J_k^{-1}(v_k - (A\hat{x}_{k-1}+BU_{k-1})), v_k - C(A\hat{x}_{k-1}+BU_{k-1})]$$

$$- \sum_1^n [v_k, v_k] - \sum_1^n \log J_k \quad , \qquad (5.6)$$

where

$$J_k = I + C(AP_{k-1}A^*)C^*$$

$$\hat{x}_k = (I - P_k C^* C)A\hat{x}_{k-1} + P_k C^*(v_k - CBU_{k-1}) \quad .$$

Maximizing (5.6) with respect to B will, of course, yield the Kalman estimate for B given $v_1, \ldots, v_n$, for the case where $\Lambda_{B_{tr}} = +\infty$, and $P_0 = \Lambda_{x_{tr}}$.

Likelihood Ratio Using Fit Error

Instead of using the Innovation we can use the Fit Error. (Cf. Section 4.1.) Then the likelihood ratio

$L(v_1, \ldots, v_n)$

$$= \exp -\frac{1}{2} \left\{ \sum_1^n [(G_k G_k^* - C_k P_k C_k^*)^{-1}(v_k - C_k\hat{x}_k), v_k - C_k\hat{x}_k] \right.$$

$$- \sum_1^n [(G_k G_k^*)^{-1} v_k, v_k]$$

$$\left. + \sum_1^n \log |J_k| - \sum_1^n \log |G_k G_k^*| \right\} , \qquad (5.7)$$

where we may replace

$$(G_kG_k^* - C_kP_kC_k^*)^{-1}$$

by

$$(I + (G_kG_k^*)^{-1}C_kH_{k-1}C_k^*)(G_kG_k^*)^{-1}$$

and

$$J_k = (G_kG_k^*) + C_kH_{k-1}C_k^*$$

by

$$(G_kG_k^*)(I - (G_kG_k^*)^{-1}C_kP_kC_k^*)^{-1} \quad .$$

We may also develop Likelihood Ratio formulas for the signal-and-noise dependent case (Section 4.6) as well as the case where the measurement noise is not white (Section 4.7), since we only need the innovation process which is defined by (4.6.3) and (4.6.14) for the former. The case where the measurement noise is <u>not</u> white would appear to be the more important one of the two in applications and therefore worth further study. In the notation of Section 4.7, the innovation process is defined by:

$$\begin{aligned} \nu_n &= w_n - \bar{C}_n\bar{y}_n \\ &= v_n - Q_{n-1}v_{n-1} - \bar{C}_n\hat{x}_{n-1} - C_nU_{n-1} \quad , \end{aligned}$$

where the recursion formula for $\hat{x}_n$ is given by (4.7.7). Of course, the variance

$$E[\nu_n\nu_n^*] = C_nF_{n-1}F_{n-1}^*C_n^* + \Lambda_{n-1} + \bar{C}_nH_{n-1}\bar{C}_n^* \quad ,$$

where $\{H_n\}$ is to be determined from (4.7.9a) and (4.7.9). The joint density in the "Absence of Signal" takes a different form now, since the noise is no longer white. In fact, we have

$$\eta_{n-1} = v_n - Q_{n-1} v_{n-1}$$

so that (omitting constants) joint density of $v_1, \ldots, v_n$ (signal absent)

$$= \exp -\frac{1}{2} \left\{ \sum_1^n \Lambda_{k-1}^{-1} (v_k - Q_{k-1} v_{k-1}) + \sum_1^n \log |\Lambda_{k-1}| \right\} . \qquad (5.8)$$

Hence the Likelihood Ratio

$$= \exp -\frac{1}{2} \left\{ \sum_1^n [J_k^{-1} (v_k - Q_{k-1} v_{k-1} - \bar{C}_k \hat{x}_{k-1} - C_k U_{k-1}, \; v_k - Q_{k-1} v_{k-1} - \bar{C}_k \hat{x}_{k-1} - C_k U_{k-1})] \right.$$

$$- \sum_1^n [\Lambda_{k-1}^{-1} (v_k - Q_{k-1} v_{k-1}, \; v_k - Q_{k-1} v_{k-1})]$$

$$\left. + \sum_1^n \log |J_k| - \sum_1^n \log |\Lambda_{k-1}| \right\} , \qquad (5.9)$$

where

$$J_k = C_k F_{k-1} F_{k-1}^* C_k^* + \Lambda_{k-1} + \bar{C}_k H_{k-1} \bar{C}_k^* .$$

Again it is not difficult to see that (5.9) reduces to (5.5) in case Q_k is zero.

BIBLIOGRAPHY

[1] Anderson, B.D.O., and J.B. Moore. *Optimal Filtering.* Englewood Cliffs, N.J.: Prentice-Hall, 1979.

[2] Balakrishnan, A.V. *Elements of State Space Theory of Systems.* New York Berlin Heidelberg Tokyo: Springer-Verlag, 1983.

[3] Cramer H. *Mathematical Methods of Statistics.* Princeton, N.J.: Princeton University Press, 1961.

[4] Gelb, A. (Ed.) *Applied Optimal Estimation.* Cambridge, Mass.: The M.I.T. Press, 1974.

[5] Jazwinsky, A.N. *Stochastic Processes and Filtering Theory.* New York: Academic Press, 1970.

[6] Joseph, P.D., and R.S. Bucy. *Filtering for Stochastic Processes With Application to Guidance.* New York: Interscience Publishers, 1968.

[7] Kailath, T. *Linear Systems.* Englewood Cliff, N.J.: Prentice-Hall, 1980.

[8] Kalman, R.E., and R.S. Bucy. "New Results in Linear Filtering and Prediction Theory." *Journal of Basic Engineering*, vol. 82 (1960): 35-40.

[9] Liptser, R.S., and A.N. Shiryayev. *Statistics of Random Processes*. Vols. I and II. New York Heidelberg Berlin: Springer-Verlag, 1977-78.

[10] Maybeck, P.S. *Stochastic Models, Estimation and Control*. Vol. 1. New York: Academic Press, 1979.

[11] Nahi, N.E. *Estimation Theory and Applications*. New York: John Wiley & Sons, 1969.

[12] Padulo, L., and M.A. Arbib. *System Theory*. W.B. Saunders Co., 1974.

[13] Papoulis, A. *Probability, Random Variables and Stochastic Processes*. New York: McGraw Hill, 1965.

[14] Parzen, E. *Stochastic Processes*. San-Francisco: Holden Day, 1962.

[15] Proakis, J.G. *Digital Communications*. New York: McGraw Hill, 1983.

[16] Rao, C.R. *Linear Statistical Inference and Its Applications*. New York: New York: John Wiley & Sons, 1973.

[17] Sage, A.P., and J.L. Melsa. *Estimation Theory With Applications to Communications and Control*. New York: McGraw Hill, 1971.

[18] Wiener, N. *The Fourier Integral and Certain of Its Applications*. New York: Dover Publications, 1933.

[19] Wong, E. *Introduction to Random Processes*. New York Berlin Heidelberg Tokyo, 1983.

[20] Zadeh, L., and E. Polak: *System Theory*. New York: McGraw Hill, 1969.

INDEX